FULL CIRCLE
The Moral Force of Unified Science

CURRENT TOPICS OF CONTEMPORARY THOUGHT

A series devoted to the publication of original and thought-evoking works on general topics of vital interest to philosophy, science and the humanities

Edited by **Rubin Gotesky** and **Ervin Laszlo**

Volume 1 SYSTEM, STRUCTURE, AND EXPERIENCE. Toward a Scientific Theory of Mind
Ervin Laszlo

Volume 2 VALUE THEORY IN PHILOSOPHY AND SOCIAL SCIENCE. Proceedings of the First and Second Conferences on Value Inquiry
Edited by **Ervin Laszlo** and **James B. Wilbur**

Volume 3 INTEGRATIVE PRINCIPLES OF MODERN THOUGHT
Edited by **Henry Margenau**

Volume 4 HUMAN VALUES AND NATURAL SCIENCE. Proceedings of the Third Conference on Value Inquiry
Edited by **Ervin Laszlo** and **James B. Wilbur**

Volume 5 HUMAN DIGNITY: THIS CENTURY AND THE NEXT
Edited by **Rubin Gotesky** and **Ervin Laszlo**

Volume 6 HUMAN VALUES AND THE MIND OF MAN. Proceedings of the Fourth Conference on Value Inquiry
Edited by **Ervin Laszlo** and **James B. Wilbur**

Volume 7 EVOLUTION–REVOLUTION
Edited by **Rubin Gotesky** and **Ervin Laszlo**

Volume 8 FULL CIRCLE: THE MORAL FORCE OF UNIFIED SCIENCE
Edited by **Edward Haskell**

Volume 9 UNITY THROUGH DIVERSITY (in 2 parts)
Edited by **William Gray** and **Nicholas D. Rizzo**

Volume 10 COMMUNICATION: ETHICAL AND MORAL ISSUES
Edited by **Lee Thayer**

Volume 11 SHAPING THE FUTURE—Gaston Berger and the Concept of Prospective
Edited by **Andre Cournand** and **Maurice Levy**

FULL CIRCLE

The Moral Force of Unified Science

EDWARD HASKELL, *editor*
Harold G. Cassidy, Jere W. Clark, Arthur R. Jensen

GORDON and BREACH
New York   London   Paris

Copyright © 1972 *by* Edward F. Haskell

Published in the United States by
    Gordon and Breach, Science Publishers, Inc.
    440 Park Avenue South
    New York, N.Y. 10016

Publishers office for the United Kingdom
    Gordon and Breach, Science Publishers Ltd.
    41-42 William IV Street
    London, W.C.2

Publishers office for France
    Gordon & Breach
    7-9 rue Emile Dubois
    Paris 14

Library of Congress catalog card number 72-84271. ISBN 0 677 12480 5 (*cloth*); 0 677 12485 6 (*paper*). All rights reserved. No part of this book may be reproduced or utilized in any form or by any means, electronic or mechanical, including photocopying, recording, or by any information storage and retrieval system, without permission in writing from the publishers. Printed in Great Britain.

"In the final analysis our compass must be our relationship with a central order..."

WERNER HEISENBERG Physics and Beyond, 1971

Dedicated to
ALFRED E. EMERSON
who taught me the basic principles of ecology and who —as Chairman of the University of Chicago's Interdivisional Committee for Unified Science, 1940–1943 —launched the course of events which has resulted in this book and the others which are to follow.

Contents

Dedication vi

Editors Statement: This is a Scientific Revolution viii

Chapter I
Summary of Theoretical Issues: What Generalization of Mendeleev's Periodic Table Means
Harold G. Cassidy 1

Chapter II
Generalization of the Structure of Mendeleev's Periodic Table
Edward Haskell 21

Chapter III
The Role of Unified Science in Vitalizing Research and Education
Jere W. Clark 91

Chapter IV
The Periodic Table of Human Cultures
Part 1: Anthropo-Socio-Historico-Linguistic Bases of the Periodic Table
Edward Haskell 111
Part 2: Direct Psychological and Genetic Empirical Basis of the Periodic Table
Arthur R. Jensen 156

Chapter 5
Unified Science's Moral Force
Edward Haskell 169

Glossary Index 215

The Author's 239

Editor's Acknowledgements 241

Appendix I 243

Index 251

EDITOR'S STATEMENT:
THIS IS A SCIENTIFIC REVOLUTION

Edward Haskell[1]

This symposium volume announces the results of a project begun in 1948 at the Centenary of the American Association for the Advancement of Science in Washington, D.C.[2] The Council for Unified Research and Education, established at that time, is here announcing the results of its then twenty-one year effort: *Assembly of the Sciences into a Single Discipline.*

Why have we waited to announce them all at once? "The kind of change we need probably cannot occur piecemeal," said Jerome Wiesner to C. P. Snow in a televised discussion. "It probably has to happen all at once." And Snow agreed.—The reason Wiesner gave is, that "The forces of tradition are strong and piecemeal changes tend to get changed back."[3]

To this important reason, another is added by Thomas Kuhn: Implicit in the traditional assortment of separate disciplines is the tacit assumption, the *paradigm*, that their data are at bottom structurally *diverse*. As Kuhn points out, however, "A scientific theory is declared invalid only if an alternative candidate is available. The decision to reject one paradigm is always simultaneously to accept another."[4]

The contrary hypothesis underlying our assembly of sciences is, that *All natural systems have a common underlying structure; and that the Periodic Table of Chemical elements is its special atomic case.* And, since an assembly is a single thing, it has to be displayed and examined all at once.

Extension of this Periodic table's structure to each of these parts of science involves what Willard V. Quine calls "clearing up the similarity notion" and discovering the universe's "natural kinds." And the procedure of extending to all of these kinds of theories a single and well established structure guarantees that these "branches of science would qualify as unified, or integrated into our inclusive systematization of nature . . . [for it makes] their several similarity concepts . . . *compatible;* capable of meshing . . ."⁵

We announced these results in 1969, in honor of Dimitri Ivanovich Mendeleev's Periodic Table of Chemical Elements. But the fact that this was the centenary of its announcement was our least important reason. (His famous paper was presented in his absence, due to illness, before the Russian Chemical Society in St. Petersburg on March 6, 1869, and was entitled *The Dependence Between the Properties and the Atomic Weights of the Elements.*⁶) Our primary reason for convening this (1969) symposium is that Mendeleev's insight and audacity, celebrated among chemists, is here extended to all—repeat, *all*—the sciences; and that his farthest out vision is, in this symposium, not transcended, but faithfully fulfilled:

"*It is the function of science,*" Mendeleev declared to his chemistry students, "*to discover the existence of a general reign of order in nature, and to find the causes governing this order. And this refers in equal measure to the relations of man—social and political—and to the entire universe as a whole . . .*"⁶

How happy Dimitri Ivanovich would have been if he could have heard us fulfil his intuitive prediction. However, I'm pleased to say that some of his compatriots did: The Voice of America included them with us in our Boston hall: that morning's Russian broadcast featured our symposium.

The structure of Mendeleev's "*general reign of order in the entire universe as a whole*" is represented by our fold-out at the end of the book. (If the reader stands it in clear view he will see in a few glances how simply this assembly works.)—Is not the Periodic co-ordinate system (right-hand side) with its five Periodic tables (and spaces for two more which are still missing) mapped into it— is it not also the fulfilment of the Royal Society's ambitious objective, set forth in 1663? Its first objective, as you remember, was the development of our discrete, one-field sciences, which has now been pretty well fulfilled. Its second objective was all but glossed over. It was stated in a single sentence: "The compiling of a complete system of solid philosophy."⁷ Of this system we present a model. Mendeleev's "general reign of order in the entire universe" turns out to be so constructed that our pursuit of the objectives we published in *Science* has turned out to be a model of the Royal Society's "complete system of solid philosophy."

We stated our objectives in *Science* as follows: To "advance social science through the stages of natural classification and evolution theory, into that achieved by the physical sciences, where scientific fields are connected, and science is closely linked to philosophy and technology."[2]

In fact, it appears to fulfill, model, map, complete, execute, generalize and assemble all the theories and predictions listed in the left-hand column of our fold-out chart, and many more for which there was no space. In his address as retiring President of the A.A.A.S. James B. Conant had said: "I place science within the area of accumulative knowledge," and had urged the coordination of its role in our society.[8]—The theoretical condition for doing this was, as we announced in *Science*, "The independent discoveries of parts of the same general conceptual scheme by students of plant, animal, and human coactions . . ."[9] This conceptual scheme was the old Periodic Table's underlying structure; and once these independent discoveries of it were recognized as such, they were developed into the Periodic tables represented in our wall display. The geometric representation of their common or universal characteristics, the Periodic co-ordinate system shown in its center, appears thus to be a model of Leibniz' celebrated project, the *Characteristica Universalis*.[10]

Our 1948 announcement in *Science* concluded with this conditional prediction: "Should this scheme prove to be a natural classification, it would create conditions for rapid coordination and advance of social sciences, as the Periodic Table did in Chemistry."[2] Its fulfilment will be set forth in our chapter on "The Periodic Table of Human Cultures."

The object of this expanded symposium is to submit our results to the scientific community so that it can judge whether or not our announced objective has been reached; and so that, if it has been reached, we can in the social and ecological sciences start reaping the kind of benefits which Mendeleev's Periodic Table, the first step in this process, conferred on the physical sciences a century ago. Namely, meaningful organization of masses of accumulated data, resolution of their paralyzing communication noise and fouling, predictions and discoveries of missing concepts and data, and increased capability of predicting changes of natural and psychosocial phenomena, and of inducing, modifying, or preventing them. In short, gaining at least some control over our destiny.

This book is an expansion of the Twenty-first Anniversary Symposium of the Council for Unified Research and Education. The original symposium was conducted, very appropriately, under the auspices of our sister organization, the Society for General Systems Research, of which all our participants but one are members.[11]

NOTES AND REFERENCES

1. Arranger of the symposium; and chairman of the Council for Unified Research and Education: C.U.R.E., Inc.
2. "Symposium on Cooperation and Conflict Among Living Organisms," *Science*, Sept 3, 1948, p. 263.
3. Program on National Educational TV chaired by Eleanor Roosevelt, *ca.* 1964.
4. Thomas S. Kuhn, *The Structure of Scientific Revolutions*, University of Chicago Press, Chicago, Illinois, 1962 (p. 77).
5. W. V. Quine, *Ontological Relativity and other essays*, Columbia Univ. Press, New York, 1969, p. 138.
6. Daniel Q. Posin, *Mendeleyev The Story of a Great Scientist*, p. 167. McGraw-Hill, New York, 1948.
   The three Mendeleyev quotations in our symposium wall display appear in the removable wall chart at the end of this book.
7. Sir Henry Lyons, *The Royal Society, 1600–1940—A History of its Administration Under its Charter*, Cambridge University Press, Cambridge 1944.
8. James B. Conant, "The Role of Science in Our Unique Society." Address of the retiring president of the A.A.A.S., Chicago, Dec. 1947, *Science*, Vol. 107, Jan. 23, 1948, p. 78.
9. Edward F. Haskell with the collaboration of Burton Wade and Jerome Pergament: "*The Coaction Compass:* A General Conceptual Scheme based upon the Independent Systematizations of Coactions Among *Plants* by Gause, *Animals* by Haskell, and *Men* by Moreno, Horney, Lundberg and Others." (The symposium's mimeographed convening paper.)[2]
10. *Leibniz Selections*, P. P. Wiener, Ed., Scribners, N.Y., 1951, pp. 5, 15–25, 66, etc.
11. Program of the American Association for the Advancement of Science, Boston, Massachusetts, December 26–31, 1969 (pp. 248–9).

HAROLD G. CASSIDY: AN INTRODUCTION

Edward Haskell

Harold Cassidy has had an important role in this assembly of sciences, this scientific synthesis. His role began when he and I were undergraduates at Oberlin College in the late 'Twenties. We lived in a house called "Αρθρον (Arthron, the Joint). Each of us in this house majored in a different field. Our endless sessions of talk then, and at our Arthric reunions through the decade of graduate studies and thirty-odd years of professional work, gave each of us an interdisciplinary and constantly updated education. During these decades, each of us made available to the others most of his papers and books as they came off the press. Thus, quite inadvertently, Arthron became a think tank.[1]

What made it goal-directed was Harold Cassidy's growing concern, in the mid-Forties, with my predicament. Being a chemist, he discerned in my writings, couched though they were in what he called "incomprehensible jargon," the actual extension of the Periodic table to biology and social science. He had the patience to follow the reasoning step by step, pointing out mistakes, suggesting improvements and, in the process, mastering those parts of the technical language without which the new abstract concepts cannot be acquired.[2]

We supported each other more and more confidently through two more decades of Arthric reunions, and eventually produced a small, privately printed book, *Plain Truth—And Redirection of the Cold War*, and now the book before you. So, over the past twenty-odd years Harold Cassidy has been my constant coach, mentor, collaborator, and link with the academic community. Without him, we would not have held this symposium, and this book would not have seen the academic light of day.

But I want to say more: While my scientific writings continued in only xeroxed and offset-printed forms, Harold Cassidy (I keep repeating his full name to distinguish him from his equally illustrious literary brother Fred), Harold has published three "Two-Culture" contributions: the first, is titled *The Sciences and the Arts—A New Alliance* (Harper, 1962). Through it, scientists and humanists can recognize each others' contributions far better than they could before. The second, is titled, *Knowledge, Experience and Action—An Essay in Education* (Teachers College Press, 1969). This Essay is essential to everyone who wants to learn, and even more, to teach— Unified Science. And here, now, is the third: the bridge across which physical scientists can explore unified science's general background-theory, of which their own discipline is shown to be a special case. And so are, of course, the major theories and laws of the biological and social sciences. In reading this chapter, one recapitulates Cassidy's twenty-or-thirty year "phylogeny", from one-field specialist to generalist. Figure IV-11 is a diagram of what this statement means.

NOTES

1. Other Arthrites are: Frederick G. Cassidy, Professor of English Literature, University of Wisconsin, Madison, Wisconsin; George T. Lodge, Professor of Psychology, Old Dominion University, Norfolk, Virginia: and Willard V. Quine, Professor of Philosophy, Harvard University, whose clarification of ontological relativity informs the concluding section of this book.
2. Hence the Glossary at the rear.

Chapter I

Summary of Theoretical Issues: What Generalization of Mendeleev's Periodic Table Means

HAROLD G. CASSIDY

1. INTRODUCTION

I would like, first, to summarize briefly the theoretical ideas present in Haskell's work as I have observed them appear and develop over the last twenty years to their present fruition. Then I would like to suggest in summary their meaning for the field of Education.

In my opinion, Haskell has discovered a scientifically-based pattern of a universal kind which is displayed in some respect by all of human knowledge and experience of Nature and Man. This is a large statement. Propositions of this kind have been advanced since the earliest days of philosophy, and in view of the signal lack of agreement among philosophers throughout the ages and today, it behooves us to be extremely wary of such statements. Yet strange things have been happening in science; and if I say that, in my opinion, this pattern that Haskell has discovered (and such discovery inevitably involves a degree of creative invention) constitutes an *invariant-relation* that enables translation between various developing fields of knowledge and experience, then at least metaphorically one can understand me to mean that like the Lorentz Transformations it makes the applicable relativity tolerable.

I assume then, that my task is to summarize the theoretical and empirical bases of this statement. That is, to give support to the hypothesis that Haskell has here a universal pattern; to show its nature and empirical reference; and to make it plausible for scientists to give it their attention. The pattern with which we are concerned is made up of several sub-patterns. I shall summarize each of these, as I see them, then put the whole together.

2. PERIODICITY

We are celebrating the centenary of Mendeleev's Periodic classification of the chemical elements. I need not remind you in detail of this work: you have before you [in the fold-out chart at the rear] a modern version of part of his Table. Essentially, what he did was to recognize a key variable, by classifying which other properties of the things classified fell into orderly patterns. This variable was atomic weight, and as you know it has been replaced by the more operationally constant property atomic number. Periodicity is displayed by the properties of the chemical elements when the elements are arranged according to increasing atomic number—and incidentally the evolutionary sequence of the production of elements must also display such periodicity. What Haskell has done is to find evidence that not only the Kingdom of Atoms, but that of Nuclei, of Plants, of Animals, of Cultures, displays a periodicity provided that the essential variables are properly chosen. This choice depends on cybernetic analysis, and its application leads directly to a sub-pattern known as 'Coaction.'

Cybernetic Basis

The essential points needed to show the relevance of a cybernetic analysis in Haskell's work can be developed with the aid of a simple diagram. Cybernetics, "the science of communication and control in the animal and the machine," as Norbert Wiener defined it in a fatherly way, deals with processes. These are analyzed in terms of the variables that change as the process occurs. That which is undergoing the process is a system.

Consider a system such as an animal organism undergoing processes which we summarize by saying it is alive. It exists in a habitat which affects it (by inputs) and which it affects through outputs. The inputs comprise work-factors such as food, air, water, and so on, depending on the organism and controls upon them, as well as contingent factors over which the animal has little control until they impinge on it. The outputs are linked retroactively by feedback to the work components. In this feedback loop, and controlling it, is the governing factor. It receives information about the state of the output, and about the state of the habitat, and by its behavior can control, through the work components, the direction of the output.

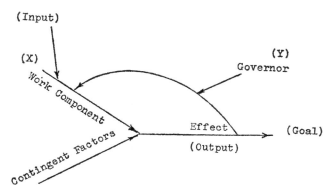

FIGURE I-1 System.

Information and control are clearly linked in the cybernetic analysis. It is the case that a system may comprise small or large groups of processes; from some ultimately single and simple process such as the reaction of two chemical substances, to the complicated, interlocked system of an unknown number of homeostatic and other processes called a living organism and, indeed, beyond to the behaviors of groups of organisms. It is the case also, that a factor that in one system is an output may exercise a governing function in an interlocking system, or may function as work component in another. This is in agreement with what we observe in the world around us.

Coaction

In the cybernetic analysis of the more complex and organized systems we recognize two distinct kinds of factors. There is the work component or components, which we shall designate X, and the governor, or controller, which we shall designate Y. Of course, the governor does work too (the strategic work), and we have simplified the relationships very greatly. There will be cases of a system made up of sub-systems, one controlling in some respects, not in others, and so on. Let us stay with the simpler case. Now, the processes that characterize X may, in the interaction with Y, be accelerated or in some way enhanced (+), or may be unaffected (O), or may be decreased (−). Similarly, the processes that Y undergoes. When the possibilities are cross-tabulated, it becomes evident that there are nine and only nine of these qualitatively different 'coactions.'

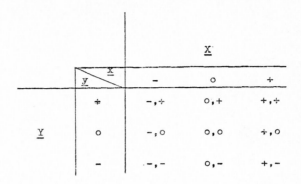

FIGURE I-2 Cross-tabulation of two coacting entities.

In (+, +) both gain. If X and Y represent two organisms, this might be called symbiosis, or mutualism. The relation (+, O) may be illustrated by the case of an older brother (Y) who, without knowing it (O), sets a constructive example (+) to a younger (X). This may be called commensalism. It is important to notice that by this classification Haskell discovered three new coactions, (O, +), (−, O), (O, −), which occur widely but had not been recognized before. To me, it is very impressive when a theory makes possible the discovery of new relationships, and especially when it completes a set of philosophical categories.

We know, of course, that there are infinite varieties of coactions, both within the qualitative differences, and quantitatively. This fact is dealt with by geometrizing the Table.

Geometric Representation

When the Table is put into a coordinate system of a special kind that you have before you, several new features appear. Eight of the coactions fall neatly upon the axes or in the quadrants, as shown[1]. The (O, O) coaction may be placed at the origin if there are no coactors. But in actual systems it must be interpreted as a third "axis," the Scalar Zero circle. This, I think, was a stroke of genius on Haskell's part. How do you decide whether a coaction has net (+) or (−) or (O) effect? You must have a reference, which is the state of the system before the coaction is initiated, or a reference point must be picked. This then establishes the (O, O) state and is,

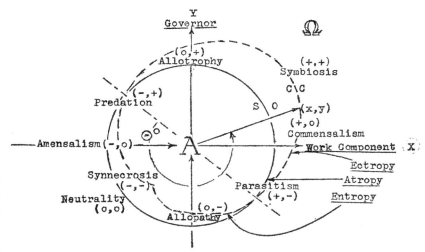

FIGURE I-3 The Periodic coordinate system: One Period.[3]

of course, neutral to all coactions. Its value is thus plotted as the radius of a circle, the Zero-Zero circle (O, O).[2]

As an example, in labor-management relations there is a profit-sharing arrangement known as the Scanlon Plan. An essential feature of the Plan is to have a reference period before it is put into operation, so that one will know whether there is actually a profit or loss under the Plan and how much it is. The value or range of a variable, measured at this time, would serve to place the (O, O) circle. From this it becomes clear that the values of net (+) coactions plot outside of the (O, O) circle in the upper right half of the manifold, and of net (−) within the reference (O, O) in the lower left. This yields a 'Coaction Cardioid.' Along the Axis of Atropy bisecting quadrants 2 and 4, the magnitudes of x and y are equal, but the signs are opposite, so the net coaction is zero. To the right and above this axis is what the philosopher Braithwaite calls the 'cooperator's surplus.' Once more we complete the philosophical categories by calling attention to the 'conflictor's deficit,' as we name it, in the lower left, net (−) part of the manifold.[3]

3. SYSTEM-HIERARCHY

A second sub-pattern is hierarchical. It is related to periodicity. The archetypal structure is that of the electron energy-level structure of atoms. Consider the structures of the first few members of the Periodic Table, arranged in the form of a System-hierarchy. Here System refers to the whole Kingdom, and the terms 'Period,' 'Stratum,' and 'Substratum' have the following significance.

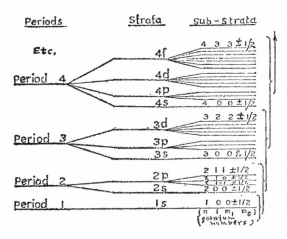

FIGURE I-4   Part of the Kingdom of Atoms, Major Stratum 2.

In this particular Kingdom, Period 1 contains one Stratum (the 'shell') and one Substratum, the orbital. Period 2 contains two Strata, the closed first shell and a second shell which shows three possible Substrata, the three orbitals. Period 3 contains three Strata, the closed first and second shells and a third, which contains five Substrata. And so on. This structure exemplifies a System-hierarchy, which has the characteristic that each higher Period comprises all previous Strata plus one new Stratum, which modifies and is modified by the others.[4] This important last characteristic is shown by Period 2, for example. The first shell of electrons is modified by the presence of increased nuclear charge and also of the second shell of electrons.

In the cases of plant and animal Kingdoms, the conventional taxonomic classification is not suitable for constructing System-hierarchies, nor for displaying periodicity. Haskell has therefore offered the additional classification scheme below which offers new insights, it seems to me. This scheme is based on cybernetic analysis. Essentially, the Periods are distinguished by a sharp change in the organism's control of its habitat. The whole scheme is based on this relationship, central to evolutionary changes in all natural Kingdoms.

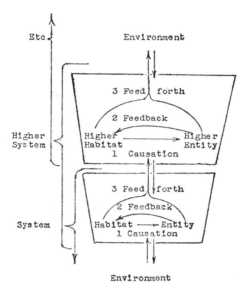

FIGURE I-5 Part of the theoretical System-hierarchy, showing the logical necessity of distinguishing *habitat* from *environment*. See Chapter II.

The steps in the hierarchy occur with changes from control-by-habitat to control-by-organism.

For example, the System-hierarchy of the Kingdom of Plants begins in Period 1 with the Protophytes and Thallophytes. These simplest living things could colonize bare rock, and convert the surface, as they grew, multiplied, and died, into a kind of 'soil.' Thus they changed their habitat. At the same time this new habitat could select for new kinds of plants. Period 2 is recognized as comprising the Bryophytes and lichens which, having roots or their equivalents, could draw nourishment from the 'soil.' At the same time they sheltered the Period 1 plants which were becoming modified too. In Period 3, the Pteridophytes, a great increase in control over the habitat came with the development of vascular tissues, stems, and leaves. Here the energy source of the sun could be more efficiently exploited, and nutriments better distributed longer distances in the plant. Finally, Period 4 plants, the seed plants, are characterized by many improvements of the habitat, including in

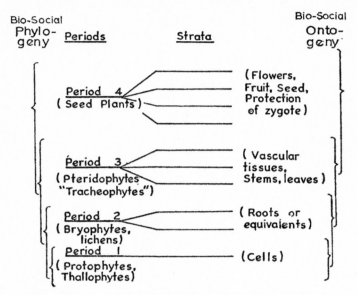

FIGURE I-6   Kingdom of plant ecosystems, Major Stratum 5.*

Editor's Note. Missing from this figure are the Sub-strata. These would, if entered, appear on the right, as they do in Figure I-4. Sub-strata represent the plant's ontogenetic stages. In accordance with Haeckel's Law, each Stratum (which developed phylogenetically) is composed of its own number of (ontogenetically developed) Sub-strata. See Chapter II. This structure may, for some people, be obscured by the fact that the controlling Stratum develops species whose adult forms occupy lower Stratum niches. Modern shrubs and grasses, for example, (Stratum 4), often occupy what used to be, in ancient times, fern niches (Stratum 3). What were once fern lands are now, in many places, grass lands. Nature, as Einstein said, is subtle. To those who see her subtlety, as Einstein did, she is not vindictive.

various cases improved selective fertilization procedures; protection of the zygote in a shell; increasing its survival chances by including with it a source of nutriment.

This additional type of classification opens new avenues for experimentation such as are suggested by terms like 'input,' 'output,' 'feedback,' and so on. At the same time it brings this Kingdom into the pattern of the Periodic Coordinate System. A similar approach deduces a System-hierarchy for Animal and one for Human ecosystems. (Chapters II and IV).

We have, then, a grand pattern which consists of the Periodic Coordinate System for the description of the state, and changes in state of systems, subsystems, and very subordinate systems, in any given Period. Each Kingdom, from Particles to Human Cultures, comprises a System–Hierarchy, each Period of which is describable in coaction terms. Superimposed on all of this is a System-Hierarchy of Kingdoms.[5] Here we are not certain about the numbering of Periods. However we do know that Particles and Atoms must have lower numbers, increasing in that order, than Molecules, Geoid Systems, Plants, Animals, and Human Cultures.

4. EVOLUTIONARY TREND

It is evident, as time has passed in the history of our world, that there has been a trend from the most chaotic state of primordial matter via a gradually habitable Earth, and the first appearance of Life, to our present state. It can be asserted, in broad terms, that two time-related phenomena are visible here. There is the Second Law of Thermodynamics which says that in any *closed* system undergoing process, a quantity called the *entropy* tends to be maximized. As far as we can judge this is a law without any exceptions—almost certainly without terrestrial exception. At the same time, we observe the appearance of organized, information- and control-employing creatures. These are always associated with *open* systems, and they decrease net entropy at the expense of their habitat. We must thus confirm, as many others have suggested, that there is also a drive in a direction which, since it has led to our existence, we can only speak of as 'upward'; that is, to increased complexity and the organisation needed to keep it viable.[6] These two directions are symbolized by Haskell in the diagrams, taking a leaf from Teilhard de Chardin, as 'Alpha,' A, and 'Omega,' Ω. They confer limits and rational bias on the entire formulation.

5. MEANING

What I wish to have made clear is that we have here an ambitious attempt at a Grand World-view. Haskell has provided us with a pattern of invariant-relations, the System-hierarchies and Periodic coordinates, which enable us to translate between all the sciences; he has provided us with a set of compatible interconnected frames of reference.

6. IMPLICATIONS FOR EDUCATION

Very basically, the essential problems in education are those of our whole culture: communication, control, and direction. Education depends in considerable part on rational discourse, though there is also a large and important non-rational component that is learned by non-verbal means: by example, and "showing how." We have seen how Haskell has discovered an overall pattern that comprises compatible frames of reference (Periodicity and Stratification) for all the sciences. This pattern, couched in operationally constant language, allows translation between the present languages of the sciences, and so should counter the present ever-increasing fragmentation among disciplines.

We see in this work the potential for improved communication, for control and direction in education. The improvement in communication comes from the use of operationally constant concepts. Words like 'conflict,' used commonly and confusingly for $(-,+)$, $(-,-)$, and $(+,-)$ coactions, can be cleared up. The people who use them can be led out of the semantical swamps in which they flounder under such usages. Precise use of words and symbols can be encouraged, and made to yield positive, constructive results.

Control is important too. To obtain an holistic education, an education that opposes fragmentation of the curriculum, and alienation and loss of integrity in the student, the education must have a rational structure. It must depend on many constraints such as always trying to tell the truth in clear and precise language. If the truth cannot be communicated in this way then the teacher must be constrained to alert the student. It should have an orderly sequence of progression that is constrained to fit the stage of physical, mental, and spiritual growth of the child. It should have an intellectual skeleton that is capable of supporting a wide variety of features as it is fleshed out with educational growth. What I am suggesting is that Haskell's world-view provides just such an intellectual skeleton. We know that the concept of coactions can be grasped by very young children—Haskell has demonstrated this. A young child can grasp, even if in a very simple way, some of his interrelations with his habitat. The thing we must be wary of, however, is the tendency in education to idolize a theory. We must engineer into any application of this work an openness to rational change, an opportunity for a variety of approaches and interpretations, and a self-revising feedback.

One implication for education that comes from this work is not only that the curriculum must be given an holistic structure, but the administrative side also. Application of systems theories is already showing remarkable results in certain school systems at the primary and secondary levels. It should be applied at the college level.

ADDENDUM

Empirical examples of the calculation of coactions may be helpful. Two aspects of the general coaction problem may be mentioned, of which the first only is dealt with here. The first is calculation from raw experimental data to a form that may be plotted in the Periodic Coordinate System. This takes care of data within Periods. But transition from one Period to another (up or down) in a System Hierarchy will probably require the use of step functions of some kind, and are not dealt with here.

We take our examples from the work of Gause[7]. Gause studied the growth behavior of protozoa, yeasts, and other organisms as they were cultured on liquid media, controlled as needed with respect to temperature, composition, pH, availability of oxygen, and accumulation of waste products. Conditions were such that the population of protozoa (or other organism) would reach a fixed level and remain there indefinitely. This was the saturation population (K) and represented a steady state with respect to the volume, or bio-mass, of protozoa present per unit volume of culture. The organisms were grown separately, and in mixture. Figure A-1 is a reproduction of Gause's Figure 10, p. 31.

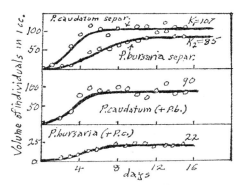

FIGURE A-1   Growth of the population of *Paramecium caudatum* and *P. bursaria* (nutrients: *S. exiguus* + *B. pyocyaneus*). From Gause.[7] Figure 10, p. 31. K is the total volume of individuals of the particular kind in 1 cc at the steady state level.

In one method of calculation we utilize Gause's K values. This gives an end-result of the processes traced in Figure A-1. One might, of course, calculate every point on the curves of mixed growth and thus trace out a line of behavior that ends at the steady-state values K (shown in Figure A-3).

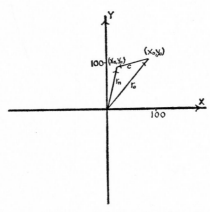

FIGURE A-2   The data of Figure A-1 are plotted to show the calculation of the coaction vector 'c'.

From Figure A-1 we see that, grown separately, *caudatum* (Y) attains the volume of individuals per cc 107. This is y_0. Similarly, *bursaria* (X), attains the value $x_0 = 85$. As a result of the interaction when the two are grown together, the steady state that is reached shows the values $y_n = 90$; $x_n = 22$. These values are plotted in Figure A-2 in the usual way. They define two vectors r_0 and r_n. The coaction vector is measured as $c = (x_n, y_n) - (x_0, y_0)$, the vector difference between r_0 and r_n. The direction of this vector shows (as the numerical data showed) that mutual depression $(-,-)$ has occurred. From the above data the components of c (a, b), may be calculated by the usual method to be $a = -63$; $b = -17$. By trigonometry, $r_0 = \sqrt{x_0^2 + y_0^2} = 136.7$; c calculates to 65.3 units of magnitude (the squares dispose of the negative signs). The angle of c calculates, from the tangent and inspection of the direction, to be 195°7' (which preserves the negative direction). This places the coaction vector in quadrant III, i.e., between 180° and 270°. We must then assign to c the negative sign: -65.3 units.

Coaction is always plotted relative to the (O, O) reference circle. This is a circle with radius r_0 (Figure A-3). The value r_0 is thought

Summary of Theoretical Issues

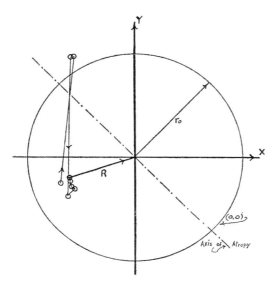

FIGURE A-3 The RO circle [reference zero, or (O, O)] of radius r_0 is shown, together with a plot of the line of behavior of the system from (x_0, y_0) to (x_n, y_n) in the phase space of Quadrant III $(-, -)$. The data are from Gause (1), Figure 10, p. 31. The wide fluctuations in the first few points are well within experimental error that goes with the presence of only a few organisms. The Axis of Atropy is shown, — . — . — .

of as a magnitude only. Moreover the symbols (O, O) are related to the symbols (+) and (−), and are not numerical zeros.

The Periodic Coordinate System is *not* Cartesian, though it does resemble the Cartesian in quadrant I. The axes are *calibrated* numerically from the origin outward. However, the X axis is interpreted to be directed from left to right; the Y from below to above. There are singularities in quadrants II and IV. These are two points which, with the origin define an axis, the Axis of Atropy, along which at any point the magnitudes of x and y are equal but their signs are opposite, so that the net coaction is zero (Figure A-3). Now, because of the defined directions of the X and Y axes, r_0 takes on the property that, on the net negative, or left-hand, part of the manifold it is directed toward the origin while in the other half it is directed away from the origin. This makes it so that if c is added to r_0 (which takes the same angle as c) the resultant vector R will fall *within* the (O, O) circle of radius r_0 when the coaction is net negative. It will extend outside of the (O, O) circle when the coaction is net positive. As we said, coaction is always relative to the (O, O) or RO circle. Thus all net positive coactions $(a + b > O)$ will lie outside of the RO circle, and all net negative coactions $(a + b < O)$ will lie inside of the circle. Where $a + b = 0$, we have the intersection of RO and the axis of atropy. This formulation causes the net plus and net minus phase spaces to stand out clearly and strikingly.

As another experimental example we take the case where *P. aurelia* grown in the presence of *P. caudatum* is found to completely displace caudatum (Figure A-4). Grown separately aurelia (Y) attains a steady state of 105 volume units per 0.5 cc; caudatum (X) under the same conditions attains 64 volume units per 0.5 cc. Grown

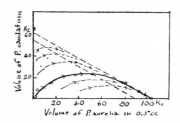

FIGURE A-4 Relative curves of coexistence of *P. caudatum* and *P. aurelia*. The solid line is from experiment; the dashed lines are from theory. *P. aurelia* eventually completely displaces *P. caudatum* under all conditions of relative abundance present at the start of the experiment. From Gause (1), Figure 4, p. 19.

together, the steady state is reached with $y_n = 105$; $x_n = 0$. Then $r_0 = 123$; $(a, b) = (-64, 0)$ and c is -64. R then is 59 at the angle 180°.

We have gathered in Table A-I data from two experiments of Gause and from two of his theoretical calculations. Because we are dealing here with different organisms, the problem of how to calibrate the X and Y axes arises between experiments and within experiments (the problem is that of 'utilities' in economics). We have arbitrarily "normalized" the data to the value $r_0 = 100$ for all the data, which is why these numbers are different from those quoted above which were taken directly from experiment. This may not be the best way to make comparisons, but it serves until a better is found. This new coordinate system cries for mathematical sophistication.

Summary of Theoretical Issues

Table A-I Data from Gause (1, 2) and calculated values of coaction

Figure	c	a	b	θ	R	Coaction
A-6	+39.2	27	28.5	46°15′	139	(+, +)
A-5	−31.0	29.6	9.85	161°35′	69	(−, +)
A-4	−52	−52	0	180°	48	(−, O)
A-3	−48	−12.4	−46	195°7′	52	(−, −)

Note. The data have all been calculated for $r_0 = 100$ units (volume of organism per cc). c represents coaction; a and b are the ordered pair that describe the vector c; θ is the angle of this vector ($\tan \theta = b/a$) relative to the X axis; R is the resultant of the vector addition $r_0 + c$; the Coactions are those in Figure 2 of this chapter.

It should be noticed that the resultant vector R *is not a coaction,* but a convenient means for visualizing the interaction. The values of R from Table A-I are plotted in Figure A-5.

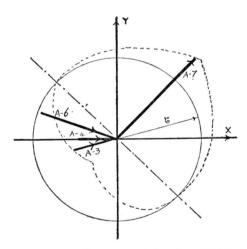

FIGURE A-5 Summary of data from Table A-I.

In summary, the coaction is calculated in a conventional way to give a coaction vector c with length $|c|$ and direction F. The sign of the value of c as determined from the relation c is (+) when $(a + b)$ [the ordered pair that describes c is (a, b)] is larger than zero; minus when smaller, and zero when a and b are equal in magnitude and opposite in sign. For a perfectly symmetrical figure, c is (+) over the range of angles from $\theta = 0°$ to $135°$ and from $315°$ to $0°$; c is (−) from $135°$ to $315°$. At angles of 135 and 315, $|c| = R$.

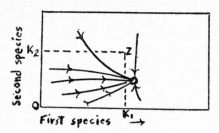

FIGURE A-6 This is the lower part of Figure 4 from Gause & Witt[8], p. 604. The coaction is mis-named commensalism by these authors. The distance from the origin to Z is our r_0; that to the focus of the curves at the steady state, is our r_n. This figure is drawn according to the theory and does not describe an experiment.

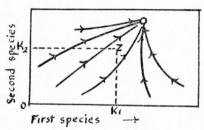

FIGURE A-7 This is the upper part of Figure 4, Gause & Witt[8], p. 604, see Figure A-6 for definition of terms.

NOTES

1. *Editor's Note.* The first quadrant is Cartesian, but the other three are not: Two of the axes are directed inward. A method of calculation is shown in the Addendum to this chapter.
2. *Editor's Note.* Also called the *Circle of Atropy*, relative to which *entropy* and its opposite, *ectropy*, can be represented and measured, as discussed below.
3. *Editor's Note.* The limits of the Periodic coordinate system are thus obviously the point of maximum entropy A and the region of maximum ectropy Ω.
4. *Editor's Note.* This Period-Stratum relation has no fundamental exception throughout unified science. Its analogous Sub-stratum-Stratum relation has an exception: its second Stratum has 3 Sub-strata (instead of 2), the third Stratum has 5 (instead of 3), the fourth has 7 (instead of 4), etc. In the atomic case, not one but two Sub-strata characterize each additional Stratum.
5. *Editor's Note.* Also called Major Periods: just as Sub-strata build up into Strata, and Strata into Periods, so Periods build up into Major Strata (Natural Kingdoms), and these build up into Major Periods (Natural Empires). The Major Periodic Table is the natural classification of Unified Science. (Chapters IV and V.)

6. *Editor's Note.* The term *ectropy* was suggested for this process by Willard V. Quine in a discussion following this symposium. Its brevity and elegance have led to its adoption throughout this book. The term *atropy* was coined by Haskell a few days later. The relations of concepts *entropy*, *atropy*, *ectropy* are shown in the Periodic coordinate system above.
7. G. F. Gause, *Vérifications Expérimentales de la théorie mathématique de la Lutte Pour la Vie.* Hermann, Paris, 1935.
8. G. F. Gause & A. A. Witt, *The American Naturalist* **69**, 596 (1935).

Chapter II

Generalization of the Structure of Mendeleev's Periodic Table

EDWARD HASKELL

1. THE STRUCTURE OF THE UNIVERSE ACCORDING TO UNIFIED SCIENCE

The universe is a Systems-Hierarchy. It has evolved in a cumulative manner, each higher step in this hierarchy, after the first, consisting of lower step components plus a new entity which has emerged out of the hierarchy, mutually modified.[1,2] The world is therefore at the same time "richly strange and deeply simple."

The objects represented by each of this dis-assembled cup's rings appear and are, of course, extremely different; yet they display a single background plan, the same when viewed from the "side" (side elevation) and from "above" (ground plan).

When you observe the cumulative edifice from below you are amazed to see that the structure of all the higher rings is potential and implicit in the forms and laws of the lower ones. And conversely, when you observe the universe from its highest rings you see that they collapse into huge numbers of their lower ring components. Theologians observing this, insist that the universe is *teleological;* that the lower kingdoms were designed to fulfil the goal of evolving into the higher kingdoms. In doing so, they arouse unending controversy with, or cold estrangement from, most one-field scientists. Assembly of the sciences, however, permits us to see that the universe is *teleomorphic* and *teleonomic*[3]; to see that the forms and laws of the lower kingdoms are such as to result in the higher kingdoms. This insight transforms the controversy and estrangement into a neutral state, where our scientific cultures' statement does not contradict the statement of our humanistic culture. The only

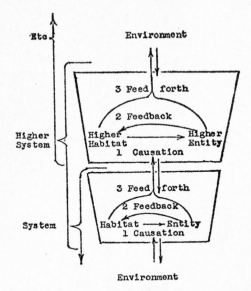

General (Abstract) System-hierarchy

The template for the assembly of empirical systems.

"A System-hierarchy is a hierarchy such, that each member of the hierarchy (except the first) consists of previous members of the hierarchy plus a new entity which the hierarchy has created, mutually modified."
Cassidy, Quine, Haskell, 1964.

FIGURE II-1 (a)

alternative to the humanists' hypothesis of Mind is the one-field scientists' equally improbable hypothesis of Chance. (It scarcely needs to be said that from the viewpoint of the theory of probability, the latter is all but self-contradictory in the immensity of its improbability.) The basis for discussion is thus strengthened, and will be expanded systematically in the following pages, as synthesis assembles the sciences into a single discipline.

Consider the deep simplicity of Unified Science: the "steps" of its great natural hierarchy fit together like the broad rings of the collapsible aluminium drinking cup, shown dis-assembled in this figure. Each broad ring represents a natural kingdom or Major Stratum. Large portions of the bottom ring, stable particles, nest

Generalization of Mendeleev's Periodic Table

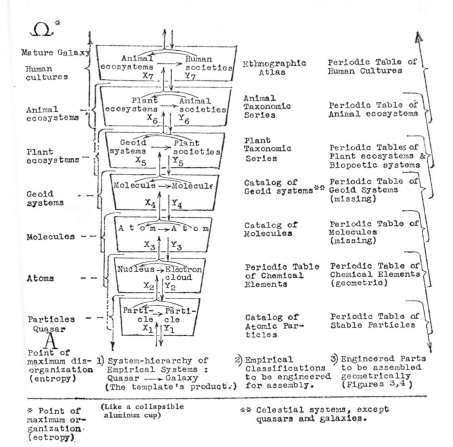

FIGURE II-1 (b)   The Cup of Life.[49]

into the second ring, atoms, as shown by means of the nested braces at the left of the drawing. (Stable particles—plus neutrons which are composed of stable particles—combine to form atoms.) Large portions of these two rings nest into the third ring, molecules. (Atoms combine to form molecules.) Large portions of these three nest into the fourth ring called geoid systems. (Particles, atoms, and molecules combine to form the lowest geoid systems, gas-dust clouds, and these form all the higher ones —stars, planets, moons and so forth.) And so on up to the highest known natural kingdom, human cultures.

The hierarchy of ecosystems extends, at the left of the drawing, from Alpha $\mathrm{\grave{A}}$ to Omega Ω, the beginning and the end—of organization. Not just the end of complexity, but of what Warren Weaver calls *organized* complexity.[4,5]

In the vicinity of Å, we represent the basic natural kingdom, that of atomic particles; the entities which, many cosmologists think, comprise the cores of quasi-stellar objects, usually called *quasars*.[6] The quasar, the nucleus of an emergent galaxy, is the habitat of atomic particles, where *habitat* is defined as *All things that an entity affects, and which affect it.*[7] This nucleus is the first ring in the Grand System-hierarchy.

Out of the quasar expand its hollow spherical shells. In the first shell some atomic particles (protons and electrons) combine to form the simplest, smallest atoms; representatives of the second natural kingdom, namely atoms of hydrogen and helium, with a single electron shell. As the first quasar shell expands, some of the small atoms build up into larger atoms, with two and three nuclear and electron-shells, while a small new quasar shell emerges inside the first. And as the large first quasar shell grows still larger, still larger atoms form, with four nuclear and electron-shells. Iron atoms have been observed in this quasar shell, indicating the existence of about a quarter of the population of chemical elements.[8] The atoms, however, *consist* of particles, and particles move into, out of, and among these atoms. At this period of its development, this quasar shell thus constitutes an ecosystem made up of two natural kingdoms: a lower kingdom of particles and a higher kingdom of atoms, always and necessarily integrated.

This System-hierarchic integration obliges us to recognize and coin a cumulative concept; one visually represented by the nest formed by the first two braces in our side elevation (Figure 1) and geometric ground plans (Figure 7). Namely, the concept, *natural empire* or *Major Period*. A Major Period is a cumulation of natural kingdoms or Major Strata.

As in human empires, so also in natural empires, the highest Stratum confers its name upon the empire as a whole. (For instance, the Roman Empire.) The kingdom of atoms being here the controlling Major Stratum, we call the second Major Period the empire of atoms.[9]

At some point in the outermost quasar shell's development—quite probably an early one—some atoms combine to form molecules, beginning the third natural kingdom and thereby, the third natural empire.

Atoms and molecules now draw together into clouds of gas, mist and dust. These are the simplest entities of the fourth natural

Generalization of Mendeleev's Periodic Table

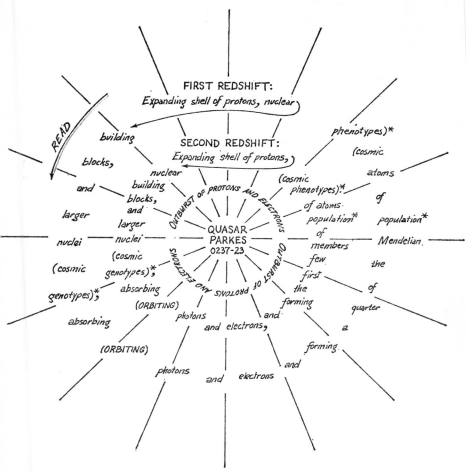

FIGURE II-2 The Genesis of the Atom Population. Reference to cosmic genotypes and the Mendelian population of atoms (cosmic phenotypes) refers to the discovery of Mendelian ratios in the atom population, and of the mechanism which probably produces them. The Chi Square test yields just under 98% significance probability[28]. See fold-out chart, top of the middle column. Data from Greenstein & Schmidt[8].

kingdom, for which we propose the name *geoid systems*. These clouds draw together by gravitational attraction, forming the rest of this natural empire: stars of all magnitudes, solar systems, meteors, and so forth. Geoid systems thus comprise all celestial bodies except quasars, out of which they emerged.

FIGURE II-3 A spiral galaxy like our Milky Way.

I now suggest that as these celestial bodies form, they pull the outer quasar shell apart along half of its equator, shrinking the great sheet of stars into the curving equatorial arm of a lens-shaped spiral galaxy; a galaxy like our Milky Way in whose oldest and outermost arm our Solar System is located.

Our Solar System's habitat is our galaxy; and our galaxy's habitat is the millions of other galaxies, jointly called the *Universe*. That may be the ultimate ecosystem, the vastest space-time system known and knowable to Man; the physical pedestal from and within which, two or three billion years ago, emerged the biosphere.[10] The fourth Major Stratum, that of geoid systems, controls the fourth Major Period represented in Figure 1 by the first four nested braces.[11]

The fifth Major Stratum emerges at a particular state in the development of a particular kind of planet in a particular kind of solar system by a richly strange process called biopoesis.[12] This process, which probably is itself Periodic[13], culminates in a tiny self-replicating sun-energy using structure called a bacterium or a one-cell plant: an entity so small that it is invisible to the naked eye. This is the emergence of the kingdom of plants, the fifth Major Stratum.

Self-replication being, however, an *exponentially* cumulative process; and being augmented by positive, self-intensifying entity-habitat retroaction, this planet's kind of surface—solid, liquid and gaseous—is modified ever more rapidly. The natural kingdom of plants herewith presently becomes the controlling Major Stratum of the fifth Major Period, the natural empire of plant ecosystems.

Let me interpose this paragraph parenthetically: At this point, which is the emergence of life, and from here on, the discussion between theologians and scientists tends to heat up or to get broken off. As you will soon see, however, the first four (non-living, abiotic or azoic) natural kingdoms are just as richly strange as the three highest kingdoms, Man's included. What fouled up the communications and parted the ways of the West's Two Cultures was not the complexity of their subject matter, nor even the scientists' shift from the authority of sacred writings to the authority of empirical data. *What fouled communications most was the scientists' long failure to assemble their findings.* How can you compare a coherently working system, such as each of the Great Religions is to its professors, with what William McElroy, Chairman of the National Science Foundation, calls "the assemblage of institutional and disciplinary fragments it (science) is largely today?"[14] However, when the sub-assemblies have been organized in a scientifically acceptable way, as I hope they are here, comparison with the theological systems becomes possible. That is why we find ourselves in an entirely new situation; one in which science comes full circle: becomes normative, and can be compared with theology; a situation in which the Two Cultures come together.

Resuming our discussion of the Systems-hierarchy: The sixth natural kingdom, that of animal ecosystems, emerged within the plant empire. But only the highest animals (which are, from the viewpoint of systems-theory, not apes but beavers) began to exert control over their ecosystems, and thus over a natural empire.

The minimal hypothesis necessary to explain the emergence of the simplest animal requires four simultaneous, coordinated changes, the kind of coincidence at which thermodynamics boggles: At some point, O. R. Anderson believes, a one-cell plant mutated part of its photosynthesizing process into rudimentary vision.[15] (It had already mutated organs of auto-motion, possibly flagelli, like those of *Euglena viridis.*) It sharpened a taste organ into an organ of smell, permitting it to follow waterborn gradients of decaying particles to their plant sources. At the same time one of its absorbing organs adapted itself to ingest plant material. And the development of genetic or non-genetic organs for information storage and association permitted, besides location and choice of distant foods and congenial habitat conditions, avoidance of dangers. This was the emergence of the kingdom of animals and of the highest kind of organization, which Teilhard de Chardin has called the *noösphere*.[16] This is the sixth Major Stratum, the highest one of Major Period 6. It emerged from the lowest Period in the Periodic Table of plant ecosystems, and only the highest animal Period in the Periodic Table of animal ecosystems, that of beaver and certain ants, achieved significant control or empire over plant ecosystems.

Beavers are true agriculturalists: Through their technology of pond and canal construction, they grow thousands of acres of plants, aquatic and terrestrial. (They regularly harvest floating algae, called *pond scum*.) Certain ants, on their part, are both agriculturalists and pastoralists. But they control small plants and animals by means of techno*geny*: elaborate genetically evolved and transmitted techniques, without which they could not live. Technology and technogeny are diverse methods by which animals achieve control over their habitats.

The seventh natural kingdom, that of human cultures, emerged from the next-to-the-highest animal ecosystem, that of the anthropoids. The first three human Periods—Lower Hunters; Higher Hunters and Lower Agriculturalists; Middle Agriculturalists and Lower Pastoralists[17,18] did not equal the beavers, control-wise. But the Higher Agriculturalists—who developed irrigation, crop-rotation and fertilization—began to do so. And the Literates, Toynbee's great civilizations, far surpassed them.[19] Our Lower Industrial civilization comprizes the highest known Period of the highest known Major Stratum of the highest known Major Period of the Universe; namely, the natural empire of Man.—Einstein affirmed that "God is a tendency in the universe."

Thus far however mankind, having developed vast technology but failed to achieve coordination of its understanding, is destroying its natural empire, *which of course includes itself*. Mankind is daily accelerating disintegration, the increase of entropy, the downward tendency toward Alpha, the collapse of the Cup of Life's highest ring into vast numbers of lower ring components, and of these components' arrangements into ever less organized states.

This is far easier to see than the opposite tendency, ectropy, for it not only happens faster but is immensely easier to produce: An animal can easily be killed, a forest fire made to wreck an ecosystem, H-bombs to destroy continents. All we have then is ashes and rocks; molecules, atoms, particles—only the Universe's lower staircases. "The highway to destruction is broad," say theologians, and they call this tendency *God's Adversary*, the *Destroyer*. Its theoretical limit we call Alpha Å. (See Figure 1, the left-hand column.)

What needs to happen for any given natural empire to be transmuted into the next lower Major Period? Simply removal of its top Major Stratum. A glance at our side elevation, Figure 1, will show that if Mankind, Major Stratum 7, is removed, we have the natural empire of animals, Major Period 6. If Major Stratum 6 is removed, we have some fragments of the natural empire of plants, Major Period 5. And so on down to Major Period 00 which is the Void.

How devolution and evolution come about, the qualitative or moral relations which produce them, whatever the natural empire, will be described two sections later on.—Summed up in a simple graph, the *quantitative* aspect of the Systems-Hierarchy appears as follows:

The principle of cumulative emergence, and of the cumulative-systemic thinking required to represent it and control it, permits us both to keep our lower mental systems and to generate higher ones: The first natural kingdom that emerges, particles, persists as an

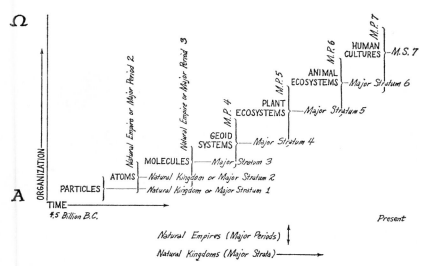

FIGURE II-4 Organization—Time Graph of the System-Hierarchy: **Natural Kingdoms combined into Natural Empires.**

entity to the present time; much of it does. A part of it, however, is organized into the next higher kingdom, atoms; a part of them both into the third; a part of all three into the fourth, and so on to the seventh, human cultures. Thus, while higher and higher natural empires emerge out of the lower, examples of the lower ones none the less persist. Thus does the span of the Systems-Hierarchy increase toward Ω. And by the reverse process, that of *removing* any given natural empire's top Stratum, the span of the Systems-Hierarchy decreases toward Å.—There is a visible correspondence between the structure of this natural hierarchy and one in human thought: the Natural number system, especially as expressed in Roman numerals: I, II, III; X, XX, XXX, etc.

2. THE PERIODIC TABLE OF CHEMICAL ELEMENTS AND ITS UNDERLYING QUANTITATIVE PRINCIPLES

It follows logically from our definition of the System hierarchy that examination of any of the natural kingdoms or Major Strata should show the general structure which all of the natural kingdoms are postulated to have in common; and that all Periodic Tables should be equally good demonstration models. (This will some day, I believe, prove to be factually true.) Today, however, at this particular point in the history of science, there is one Major Stratum, that of atoms, whose Periodic Table has been worked out and verified incomparably further and better than any other. This Periodic Table, therefore, is the classification whose structure has been generalized.—First, let us consider this model in its traditional form; the form based on empirical data; then in its geometric form obtained by generalizing its structure Systems-Theoretically: the Periodic Coordinate System.

The Periodic Chart of the Chemical Elements appears on the walls of chemistry labs and class rooms around the world; and it appears in several forms. Whatever its form, it has developed from a brilliant arrangement of data about the chemical elements which early 19th century chemists and physicists had accumulated. This classification was announced in 1869 by the Russian chemist Dimitri Ivanovich Mendeleev.[20] Its structure has been verified, extended and improved ever since, and currently displays the form given it by Glenn T. Seaborg, Chairman-emeritus of the U.S. Atomic Energy Commission.[21] For purposes of exposition, however,

Generalization of Mendeleev's Periodic Table 31

I have simplified its representation into a form closer to the one originally proposed by Mendeleev.[22]

All of the hundred-odd sets of atoms called *chemical elements* fall empirically into nine Groups; the nine patterns into which the major properties of the chemical elements resolve themselves. The Groups appear as the vertical columns numbered from I to VIII, plus the right-hand column which may conveniently be called Group O. And, since there are over a hundred of these sets, these patterns are repeated Periodically. These Periods form the table's horizontal rows.

Mendeleev did not try to explain theoretically why there are nine Groups. (As a matter of fact, when he first announced his Periodic Table, it had just eight: the Group O elements, the inert gases,

MENDELEYEV'S PERIODIC CHART OF THE ELEMENTS

group→ per↓d	I	II	III	IV	V	VI	VII		VIII			O
	Fr	Ra	Ac									
	Au	Hg	Tl	Pb	Bi	Po	At					Rn
	Cs	Ba	La	Hf	Ta	W	Re	Os	Ir	Pt		
	Ag	Cd	In	Sn	Sb	Te	I					Xe
	Rb	Sr	Y	Zr	Nb	Mo	Tc	Ru	Rh	Pd		
	Cu	Zn	Ga	Ge	As	Se	Br					Kr
	K	Ca	Sc	Ti	V	Cr	Mn	Fe	Co	Ni		
	Na	Mg	Al	Si	P	S	Cl					A
	Li	Be	B	C	N	O	F					Ne
							H					He

MODERN FORM OF SPECIAL CASE (ATOMIC) 1869

FIGURE II-5a The Kingdom of Atoms: Major Stratum 2. The empirical classification from which the Periodic coordinate system was derived.

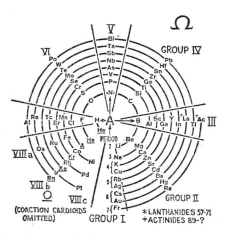

FIGURE II-5b The Periodic Table of Chemical Elements mapped into the Periodic Coordinate system. Simplified: coaction cardioids omitted. (Shown in Figure 11.)

were all discovered later.) The reason why there are nine Groups appears to be systems-theoretic in nature. It holds good for all the Periodic Tables to which the structure of Mendeleev's table has been generalized, and is one of the principal hypotheses upon which this assembly of the sciences is based.

The systems-theoretic reason why there are nine Groups is as follows: Natural systems are basically made up of two coacting components, shown in the general or abstract concepts diagrammed in the central part of Figure 1a: The *habitat* or *work component*—which cyberneticians often call collectively *the factors*[23]—and its usually much smaller *entity*, which they call the *controller*. (In atoms, the controller is the small, massive nucleus; the work component is the large electron cloud.) Those properties of the system that fall into the Groups, manifest the theoretically and empirically possible sets of coactions between the work component (habitat) and the controller (entity).

The relations between these two basic systems components are fundamentally action (or causation) and retroaction which, when it activates the controller, is called *feedback*, shown as a leftward arrow in Figure 1a. These cybernetic relations are jointly called *coactions*,[24] and are obtained by considering the totality of rate-changes which each coactor can induce in the other. Not only do + and − represent rate changes (acceleration and deceleration, respectively) but 0 too is regarded as a rate-change; namely, as 0 rate-change. The theoretical and empirical totality of coaction is then obtained by cross-tabulating +, 0 and − for work-component and controller, as in Figure II-6, top.

3. THE PERIODIC COORDINATE SYSTEM

This cross-table Gestalt changes almost at once into the conventional pattern that every school child learns, the plane Cartesian coordinate system shown in the center of Figure 6. It changes into a Cartesian coordinate system on whose X axis Unified Science maps the system's work component, and on whose Y axis, the controller.

The Cartesian coordinate system, however, has just one point at which coaction is (0, 0), its center or origin. It follows that the rate-changes of only one kind of system at a time can be mapped into it. This means that it cannot be the framework into which the sciences are to be assembled. To unify the sciences—to map the Periodic Table of chemical elements geometrically, as in the lower part of Figure 5, and all the other Periodic tables representing the empirical Systems-hierarchy from Å to Ω into a *single* coordinate

Generalization of Mendeleev's Periodic Table

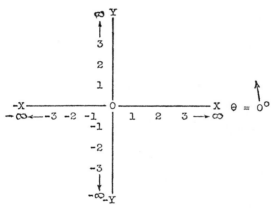

FIGURE II-6a   The Coaction Cross-table.

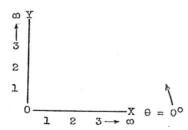

FIGURE II-6b   Newton's Cartesian Coordinate System: Zero, positive, and negative numbers. (Aberration from Descartes' system.)

FIGURE II-6c   Descartes' Coordinate System: Zero and positive numbers only. Completion of this coordinate system yields the Periodic coordinate system, Figures 7, 8, 9, 10.

system (below)—a drastically new coordinate system had to be invented, and a suitable method of calculation worked out for it.

To show how it came about—and how it involves a change from the traditional, mutually exclusive categories to inclusive, nested categories—this invention will be considered in the context of the history of the coordinate system:

The first quadrant of the traditional (Cartesian) coordinate system (Figure 6, center) was the first to be invented, and remains firmly in place in the Periodic coordinate system. (It was invented in Ancient Egypt and reinvented in ancient Greece.) Why only quadrant 1? Because these civilizations had only the Natural number system, which at that time started with the number 1 and has just positive integers.

This one-quadrant coordinate system was lost during Egypt's dark ages, was reinvented in Greece, and lost again in the dark ages resulting from the disintegration of the Graeco-Roman civilization. Some time after the rise of Western Christendom it was reinvented independently by Descartes and Laplace in a more advanced form, one which includes zero, as shown in the bottom part of Figure 6.

It was not a complete concept. However, two of Descartes' famous dreams of the night of March 19, 1619 in Nüremberg indicate that its completion would have to result, not in the Cartesian, but in the Periodic coordinate system: Descartes' principal dream was as follows: "I saw physics reduced to geometry and all the sciences extending out of it as a great chain".[25,26] This chain seems to appear in Figure II-1b as the interlocked braces whose first four links represent the physical sciences, and out of which extend three more links representing the biological and social sciences. (These linked braces reappear much more graphically, however, in Descartes' quadrant of the Periodic coordinate system shown in Chapter V.)

The other of Descartes' obviously relevant dreams was simply "*Est et non*" ("Is and not"). This can be interpreted as positive numbers (*est*) and zero (*non*), and thus as implying exclusion of negative numbers. This exclusion is not trivial: it is essential to the realization of Descartes' first dream, the great chain of geometrized sciences.

To achieve this ancient objective, it is obviously necessary to start the coordinate system with the symbol which represents the very bottom of the Systems hierarchy. This symbol is Alpha Ă which represents the point of maximum disorganization or entropy. It lies at the origin, the center, of the Periodic coordinate system, Figure II-7.

This part of unified science's coordinate system may be regarded as the plan view of the "Cup of Life," whose side elevation appears in Figure 1b: the seven Major Periods are here shown in their nested position, viewed from "above."

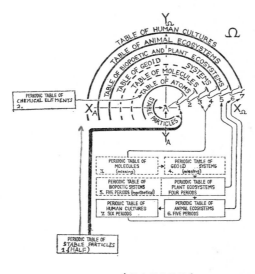

FIGURE II-7 Periodic Coordinate System: zero and positive numbers. The Major Strata are shown, each representing a Periodic Table. Matter and Mind.

Let us pass lightly, for the moment, over the half-circle near A representing half of the kingdom of stable particles. (The other half, representing the stable anti-particles, is shown in the Inverted Periodic coordinate system, near the center of Figure II-8.)

We thus come to the second Major Stratum, the kingdom of atoms, represented by what will be called a Major scalar-zero circle. This circle represents the whole geometrized Periodic Table of chemical elements shown in Figure II-5. Each of the other five nested scalar-zero circles in Figure 7 similarly represents an entire geometrized Periodic table, though they cannot be shown in this short book.[27] Each of the natural kingdoms or Major Strata is represented by a nested set of Periods, like the one representing Major Stratum 2, the kingdom of atoms (Figure 5.) Thus, each of the broad rings of the "Cup of Life," shown in Figure 1, turns out on closer inspection to be itself a "collapsible drinking cup." Whether Major Strata 3 and 4 (the kingdom of molecules and the kingdom of geoid systems) display Periods, and if so how many, remains to be determined. But the Kingdom of plants (Major Stratum 5) appears to have four Periods, the kingdom of animals appears to have five, and the kingdom of human cultures (which we will discuss in Chapter IV) seems to have six Periods, and to be in process of generating a seventh.[28]

The chain of interlocked braces in the Periodic coordinate system appears to be Descartes' "great chain of sciences." It is inherent in this model of unified science, and had been drawn some years before I read of Descartes' dreams. I am, however, most happy to acknowledge the priority of his two interdependent discoveries: limitation of the coordinate system to positive numbers and 0 (*Est et non*), and the nesting of seven Periodic coordinate systems which this makes possible, and is represented by his "great chain of sciences."[29] (To interpret Descartes' 17th century dreams in this meaningful way it was, I think, necessary to have executed them previously in 20th century detail.)

This new detail, moreover, prevents us from omitting the breakdowns of higher systems into their lower system components. Since evolutions of higher systems are here represented as extending outward from Å along the X_Ω and Y_Ω axes, their breakdowns or *de*volutions have to be represented as progressing inward toward Å, and to be mapped along the inward-directed X_A and Y_A axes. (See Figure 7.) This mapping does not, however, imply the use of negative numbers. What it implies is subtraction from the upper limit of this coordinate system: from its largest zero circle, with radius Omega, (*Est*), toward its lower limit Å (*Non*). (Also subtraction within each Period, from its (0, 0) circle, as will be shown.)

Mathematicians, of course, see instantly that the method of calculation appropriate to the Cartesian coordinate system does not suffice for the Periodic coordinate system. The methods appropriate to the latter *include* the former, but go beyond it.—One method, developed by Harold G. Cassidy, is set forth in the Addendum to his Chapter I. Another method, which is still incomplete, was developed by Gause and Witt, following Volterra, Lotka and others.[30,31] At the suggestion of G. Evelyn Hutchinson, I generalized this construct in 1947, thereby obtaining the Periodic coordinate system. However, while their equation works in quadrants 1 and 3, Cassidy and I have not succeeded in adapting it to quadrants 2 and 4. Through their method is promising, it is therefore here omitted.—Interested mathematicians are invited to take up this calculation tool and make it useable throughout the Periodic coordinate system. It can almost certainly be done.

Turning now to the part of Unified Science's Coordinate system which hardly anyone but particle physicists and astronomers

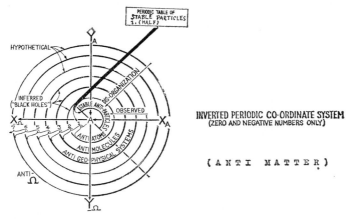

FIGURE II-8   Inverted Periodic Coordinate System: zero and negative numbers. Anti-matter only; Major Strata.

habitually think about, it scarcely needs to be pointed out that for the representation of the probably immense regions of the universe inhabited by anti-matter, exclusive use of zero and negative numbers is just as essential as exclusive use of zero and positive numbers is for the representation of our familiar world of matter. There are so-called "black holes" in the sky: great celestial bodies through which matter apparently converts itself into anti-matter,[32] presumably generating an Anti-System-hierarchy from Å to Anti Ω. The Inverted Periodic coordinate system, into which anti-atoms and the rest of the Anti-System-hierarchy can be mapped, appears in Figure II-8.

Some effects of the anti-particles have been observed. What they imply has, theoretically, to be extrapolated into the construct of an anti-universe, as represented in this figure. Our understanding of the nature of anti-matter implies, however, that partially material beings such as we are cannot verify this figure's hypothesis empirically: To do so would annihilate our material parts. For this very reason however, it is important to consider the probably immense anti-universe scientifically. And the Inverted Periodic coordinate system constitutes, it is hoped, a contribution to this study.

In any case, it is essential to unification of the sciences. For the first natural kingdom of the anti-universe, the stable anti-particles, constitutes just under half of the single Period which comprises the Periodic Table of stable particles, Figure 9. This is the Major

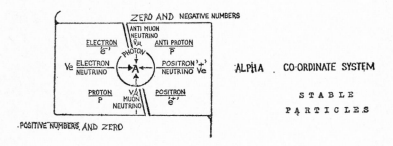

FIGURE II-9 Alpha Coordinate System: zero, positive, and negative numbers. (Inverted Cartesian Coordinate System, Figure 6b). Stable Particles.

Period which the universe and the anti-universe have in common; and its representation is the bridge linking the Periodic and the Inverted Periodic coordinate systems to each other.

The Alpha coordinate system's structure is, on one hand, implicit in the structures of the two Major Periodic Coordinate Systems; and on the other hand, determined by the data mapped into their lower parts. The transitional coordinate system consists of the other two coordinate systems' inward-directed axes. It is thus a combination (and repetition) of the end segments of these four axes, and turns out to be the Cartesian coordinate system with axes and coordinates reversed. For obvious reasons I have called it the Alpha coordinate system.

Mapped into it is the Periodic Table of stable particles: This is obtained first by repeating the half-circles near the origins of the Periodic and Inverted Periodic Coordinate Systems and then by adding the photon which, being its own anti-particle, maps into its scalar zero circle. To correspond with reality, the Periodic Tables of all the material natural kingdoms beyond particles must be confined to zero and positive numbers; those of all anti-material higher kingdoms, to zero and negative numbers, as we have seen.

Relative Plus and Minus as well as Zero

There are different kinds of pluses and minuses: absolute ones, representing pro-matter and anti-matter which annihilate each other; and relative ones representing, say, proton (+) and electron (−), whose mutual attraction is essential to the formation of matter.[33] It thus turns out that the symbols (+) and (−) have long represented quite different kinds of things, just as has (0).

Long ago Kelvin discovered a radically different kind of zero than Celsius' relative zero, the freezing point of water at sea level; namely cessation of translational movement, which is absolute. What we now find are fully as different kinds of + and −. The difference between protons +P and anti-protons −P, and between electrons −e and positrons +e is absolute: They literally nullify each other. But the difference between protons +P and electrons '−'e is relative: they combine to form atoms just as −P and '+'e probably combine to form anti-atoms. Only when these distinctions are mirrored in our notation is it possible to map the stable particles into a coordinate system. This mapping appears as shown in Figure II-9, where semi-quotes distinguish relative plus ('+') and relative minus ('−').

The four neutrinos fall on the four coordinate axes, each of whose coaction symbols contains a zero (Figure 6). Protons and anti-protons fall in quadrants 3 and 1, respectively. Their coaction symbols (+, +) and (−, −) nullify each other, producing (0, 0) as the corresponding natural phenomena "annihilate" each other, producing photons (0, 0).

Similarly, electrons (−e) and positrons (+e) fall into quadrants 2 and 4, respectively. Electrons are labeled '−'e because they and protons (+P) attract each other. They form atoms, systems *higher* in the hierarchy, which is the opposite of what positive and negative numbers do to each other.

It is thus clear that all minus quantities in the Periodic coordinate system (Figure 7) are relative: They represent subtraction from some *relative*-zero circle. It thus turns out that there need to be relative and absolute minuses and relative and absolute pluses just as there are relative and absolute zeros.

I present the Alpha Coordinate System in response to the following hope voiced by the nuclear physicist C. S. Wu: "Physicists," she wrote, "continue to study the smallest fragments of matter in the hope that careful analysis will fit quantitatively with a simple mathematical pattern that is yet to be found."[34] The stable particles constitute but a small part of the kingdom of particles. Their mapping into the Alpha Coordinate System may, none the less, prove to be a step in the direction called for by Dr. Wu.

These three frames of reference—the Periodic, the Alpha, and the Inverted Periodic coordinate systems—comprise a single, logically consistent system whose components scientists can handle individually. Mathematicians can, however, visualize them as a single hyper-spatial coordinate system.[35] In either case the coordinate system of unified science by-passes (or, rather, incorporates) the four-quadrant construct shown in the center of Figure 6, the Cartesian coordinate system.

This most used of all coordinate systems, was designed by Isaac Newton some decades after Descartes' famous dreams. It incorporates Descartes' one-quadrant coordinate system (bottom of Figure 6). But, by violating Descartes' dream *Est et non* (through inclusion of negative numbers), it precludes the realization of Descartes' dream of geometric unification of the sciences. This is the fateful rock which shattered the stream of science's development into a fan of *separate* disciplines, and thereby shattered the mind of Western civilization into two mutually paralyzing parts, which C. P. Snow has aptly called the *Two Cultures*. On one hand the great historic culture of the humanists; on the other hand the fatefully discoordinated culture of one-field scientists and technologists; and between them the confused gulf of mutual dislike and non-understanding.[36]

How have we by-passed the Cartesian rock of cultural division? By making it the keystone of the unified culture which Snow predicted would presently emerge in the United States. We have inverted the Cartesian coordinate system and placed its origin or center in a position corresponding to the one displayed by Nature: the mysterious solitary point located between the objective universe of pro-matter, to which we equate the subjective universe of absolute positive numbers; and the universe of anti-matter to which we equate the mental universe of absolute negative numbers.

Generalization of Mendeleev's Periodic Table

Instead of many separate Cartesian coordinate systems, each centered on a relative-zero point which bears no relation to the other systems' origins, we now array the series of concentric, hierarchically related scalar-zero circles. The function of the Cartesian system's absolutely negative numbers which we have assigned to anti-matter is now fulfilled by the subtraction of positive numbers from these scalar-zero circles in the direction of Å.

Conclusion

There is a series of discontinuities which mark the boundaries of what Quine calls *natural kinds*.[38] The greatest of these is the boundary between pro-matter and anti-matter, which I propose to call Å. The next most important boundaries separate the natural kingdoms or Major Strata, represented by the concentric circles in Figures 7 and 8. The third most important natural kinds are those of the Periods within each Major Stratum represented, for instance, by the circles in the lower part of Figure 5 (kingdom of atoms), but implicit in each of the other Major Strata as well. The third and fourth most important natural kinds are Strata (e.g. electron shells and nuclear shells) and Sub-strata (e.g. their orbitals). These too are implicit in the Periods of all the other Major Strata.

Relative (+) and relative (−) thus play just as important roles in unified science as does relative (0), which reappears in each Period and Major Stratum.

The formal quantitative aspect of unified science is summed up by putting together the three parts of unified science's coordinate system.

Data which would heretofore have required an indefinite number of arbitrarily arrayed Cartesian coordinate systems are hereby assembled into a single hyper-spatial construct whose center turns out to be a Cartesian plane coordinate system in reverse: the Alpha coordinate system. This rock of fateful division, by-passed in its original form, thus turns up inverted and renamed at the center of the mental organization which Leibniz predicted under the name of *Universal Characteristic*.[37] This is the absolute point at which Relativity—both physical and ontological—begins, and from which they derive coherent existence and meaning in the mind of Man.

We turn now to the Universe's qualitative principles.

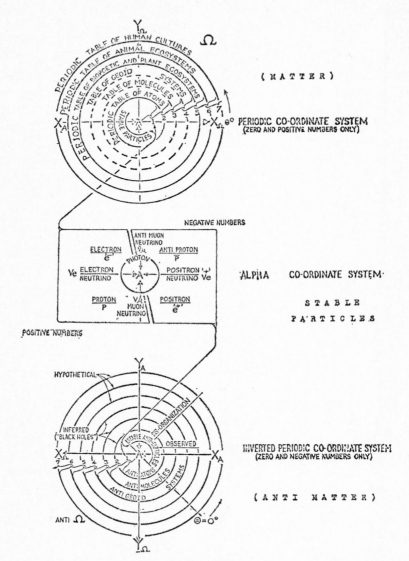

FIGURE II-10 The Coordinate System of Unified Science: Quantitative Aspect.

4. QUALITATIVE ASPECTS OF UNIFIED SCIENCE

One of the surprising ways in which this assembly of special sciences, this *whole*, exceeds the sum of its parts is the emergence within it of the Moral Law. Gottfried von Leibniz foresaw this clearly in the seventeenth century, as will be shown. And Werner Heisenberg has restated it for the twentieth century in *Physics and Beyond* as follows: "The problem of values ... concerns the compass by which we must steer our ship if we are to set a true course through life. The compass itself has been given different names by various religions and philosophies: happiness, the will of God, the meaning of life—to mention just a few ... I have the clear impression that all such formulations try to express man's relations to the central order. Of course we all know that our own reality depends on the structure of our consciousness; we can objectify no more than a small part of our world. But even when we try to probe into the subjective realm we cannot ignore the central order or look upon the forms peopling this realm as mere phantoms or accidents. ... In the final analysis, the central order, or the 'one' as it used to be called and with which we commune in the language of religion, must win out. And when people search for values, they are probably searching for the kind of actions that are in harmony with the central order, and as such are free of the confusion springing from divided, partial orders. The power of the 'one' may be gathered from the very fact that we think of the orderly as the good, and of the confused and chaotic as the bad."[39]

When the Periodic coordinate system first appeared (in 1940), I called it *The Coaction Compass*.[40] Soon after that an article appeared in the *Scientific Monthly* entitled "The Religious Force of Unified Science."[41] To whatever extent Unified Science is an assembly of the whole, it represents, of course, the "one"; to whatever extent this assembly meets the scientific tests of correctness, its structure must correspond to the central order; and to whatever degree this happens, it represents the Moral Law.

This conclusion is implicit in the Systems Hierarchy depicted in Figure 1, the "Cup of Life": The apex of this hierarchy, the human mind, Heisenberg's "subjective realm", consists of representations of the previous systems, plus something more which has emerged from this hierarchy (namely, Man's mind), mutually modified.[42]

Unification of the sciences in any human mind implies this mind's assumption of the order central to the universe. For to the extent that in this mind, accurate representations of the universe's parts (one-field sciences) are so related to each other as to yield verifiable predictions, it probably is correct. To the extent that it is correct it probably results in "the kind of actions that are in harmony with the central order . . ." And that is the Moral Law.

"We think of the orderly as the good," Heisenberg says, "and of the confused and chaotic as the bad." In less traditional terms, the increase of disorder is called *increase of entropy* and the increase of order is called *increase of ectropy*:[43] In 1947 Warren Weaver pointed out that the sciences should be arrayed relative to these two universal tendencies.[44] This has accordingly been done: the point of maximum entropy constitutes one limit of our organizing framework; the point of maximum ectropy its other limit. The whole quantitative aspect of the universe's moral manifestation, from Omega through Alpha to Anti-Omega, is shown in Figure 10. That is to say the range of one aspect of the Moral Law; its quantitative range.

These theoretical limits of entropy and ectropy constitute boundaries for mental organization, over-all. Each increase or decrease, however, implies a zero point, relative to which it has been recognized and then perhaps estimated or measured. Unorganized science permits these zero points to be arranged in *ad hoc*, hap-hazard, random fashion, as they long have been. Organization of the sciences, on the other hand, requires a rational, empirically based arrangement of zero points or lines, centered at the center of existence A, and extending to its upper limit Ω.

Since the universe's major moral tendencies are called *entropy* and *ectropy*, I propose for their zero or reference-tendency, the term *atropy*; and, as its geometric representation, a circle.

This theoretical concept may suggest what to look for empirically. What would atropic systems be? Would they not be inert systems; systems which neither combine with others ectropically, nor disintegrate other systems entropically? In the kingdom of atoms there is a complete Group of such systems: the inert or noble gases. They constitute Group 0 of the Periodic table of chemical elements, Figure 5.

How could they be represented geometrically? The limit of entropy is the point Alpha, the point where the regress of coordinate systems stops. A point may be regarded geometrically as a circle with zero diameter. Increases of ectropy may therefore be represented by concentric relative-zero circles with ever larger diameters, as

Generalization of Mendeleev's Periodic Table

shown in Figure 7; and increases of anti-ectropy may be shown in the same way, Figure 8. Organized representation of the Major Strata can thus occur in terms of concentric circles of atropy.

In geometrizing the Periodic Table of chemical elements (Figure 5, b), we begin by relating each Period to a zero circle, defined by an inert element, classed in Group 0; an element whose characteristic coaction is *atropy* (0, 0).

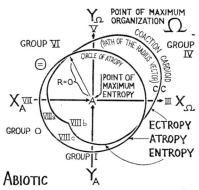

FIGURE II-11 One Period of the Periodic Coordinate System showing the circle of atropy and the coaction cardioid. Radius angles are measured counterclockwise from the X_Ω axis. They are called *theta* angles, and their symbol is θ. Hence the generalized Periodic Law reads, $R = f(\theta)$. Its graphic representation is the coaction cardioid. The Groups of the Periodic Table of chemical elements, should be mapped on cardioids.

Let us assemble the three major manifestations of the Moral Law: the non-living or *abiotic*, the living or *biotic*, and the human or *cultural*, organizing our vast warehouse of what has been called, by William Eblen, the A B C interrelationships.[45]

The early formulations displayed various aspects of the Moral Law. They were statements about cultural relations, Major Stratum 7; and were, moreover, made thousands of years ago in the idioms of simpler cultures. "As ye sow, so shall ye reap."[46] "Do unto others as you would that they did unto you."[47] It is therefore natural that to this day discussions of values are made almost exclusively in terms confined to Man; and that, even by physicists. "The problem of values ...", says Werner Heisenberg, "concerns the compass by which we must steer our ship if we are to set a true course through life." And that includes the whole Systems-hierarchy.

Unified Science agrees unequivocally that values constitute the most determining human relations. But it goes further: it points out than Man belongs to the category of Culture, the highest member of the Systems-hierarchy, and that by definition these self-same values must occur in various forms on its lower levels. Values must have been, and must still be displayed by biotic systems (by animal and plant ecosystems); and by abiotic systems: (geoid, molecule, and atomic systems). In fact, they must be displayed even by particles: Teleonomy, it was pointed out above, appears on every level of the Systems-hierarchy. It follows, of course, that we have been seeing the Moral Law at work on each of these levels all our lives without recognizing it, as people had seen falling apples and other objects all their lives without recognizing that they display the law of gravity. When each of these laws is pointed out, stated verbally, and formulated mathematically, so that it can be clearly verified or disproved, the gate is opened to our Two Cultures' complete, organic union.

The first formulation of the complete Moral Law for a non-human natural kingdom was Dimitri I. Mendeleev's 1869 announcement of the Periodic Law: "The properties of the chemical elements are functions of their atomic weights."[20]

This looks no more like a statement of the Moral Law of Man than a falling apple looks like an orbiting planet. Yet what it states for atoms is that "As ye sow, so shall ye reap," where "reaping" is the properties of the chemical elements and "sowing" is (as I will show) the coaction between the atom's work component, its vast, light electron cloud, and its controller, its tiny, massive nucleus. This "reaping" is the result of the doing of the one to the other, and the other's doing unto it.

Hindsight shows that atomic weight was the first manifestation of intra-atomic coaction discovered and measured by scientists. Its next manifestation was atomic number, the number of protons in the atom's nucleus. So the Periodic Law of atoms was improved to read "The properties of the chemical elements are functions of their atomic numbers." But with the (unconscious) discovery by T. H. Langlois[48] that *The properties of animal societies are functions of coactions between work component and controller* (the 97% of small and medium fishes in the ponds, and the 3% of the largest fishes), the Periodic Law became a black box problem.[49] As these two components of animal systems did unto one another, so they reaped:

Where the Majority and Minority cooperated (+, +), all their societies' properties were positive: the largest percentage of the populations survived, they grew the best, had the least sickness, and formed the most cohesive societies. Where the Majority and Minority persistantly damaged each other (−, −) all these properties were negative: the smallest percentage of the populations survived, they grew the least, had the most sickness, and had the most disintegrated societies.

Langlois, it is true, discovered only three or four of these coactions and their resulting property configurations in animal societies; and he did this empirically, as Mendeleev had done with atoms before him. But with the Periodic Law in existence; with the knowledge that in its atomic case there are *nine* Groups; and with Toynbee's clear demonstration in human civilizations that the corresponding properties of Man's societies are functions of the coactions between the Majority and the Minority, the black box problem of the explanation of the Periodic Law could be formulated, and a tenable solution be obtained: *If* the properties of systems in general—abiotic, biotic, and cultural—were functions of coaction between work component X and controller Y *then* their functions should be exactly *nine* Groups of properties (Figure 6); for instance, the nine Groups in the Periodic table of chemical elements, Figure 5.

The start of a tenable solution was obtained in 1940 by means of the crosstable shown in Figure 6, which yielded the requisite number and kinds of Groups. This tentative numerological approach to a solution was confirmed theoretically in 1948 by Norbert Wiener's "*Cybernetics*—Or Control and Communication in the Animal and The Machine."[50] This, and the broader Systems-Theoretic works of von Bertalanffy[51,52] and de Latil[23] back up, and articulate Mendeleev's, Langlois' and Toynbee's a, b, c discoveries theoretically, showing them to be Systems-hierarchic manifestations of one universal law: "As ye sow"—as your work component and controller "do to one another"—"so shall ye reap"—such shall be your system's properties.[53]

Over a number of subsequent years the Periodic coordinate system was developed to express general, abstract relationships jointly displayed by these empirical a, b, c phenomena. And the empirical Periodic table of chemical elements (Figure II-5a) proved mappable into it (Figure II-5b).

When the Systems-hierarchy was defined by us (1964), it implied what actually is a very old idea: that any fundamental Law which obtains in one Major Stratum of the General Periodic Table obtains in all. That a single Law pervades the universe.

This concept is implicit in Leibniz's 17th century theory that it would one day be discovered that all sciences have a common structure. This prediction appears to be fulfilled by the Periodic coordinate system, whose recurrent structure is displayed (for the Abiotic Major Strata) in Figure 11.

The Periodic coordinate system is Cartesian with respect only to quadrant 1; and even there, just in respect to the circle of atropy. This circle's radius is determined by the intersection of perpendiculars erected in the cartesian manner at coordinates located on the X_Ω and Y_Ω axes. This circle-of-reference represents the derivative of position (zero) which is the same in all quadrants, whether the axes are Cartesian or Periodic.[54]

This coordinate system is so constructed that all its higher derivatives—those of rate-change (of positive and negative acceleration) and thus the path of its radius vector—obey the General Periodic Law stated below. The path of the radius vector, therefore, turns out of the circle of atropy in the positive, Omega-ward part of the plane (in Greek, *out-turning* is *ectropy*); and it turns inside the circle of atropy in the negative, Alpha-ward part of the plane (in Greek, *in-turning* is *entropy*).

By letting the vector's scalar or length component (R) represent the system's properties, and letting the vector's directional component (θ) represent the coactions of the system's work component X and controller Y, we obtain as the radius-vector's path, the coaction cardioid or heart-shaped curve in Figure 11. This is a geometric representation of the Periodic Law in its general or universal form: the form which applies to the kingdom of atoms, as Mendeleev's law does, and to all other natural kingdoms as well. The geometric, general form of the Periodic Law thus reads as follows:
$$R = f(\theta)$$

The length of the radius vector (R) is a function of its direction (θ). This says in System-theoretic language that "The properties (R) of a system are functions of (are determined by) the coactions between its work component and controller (θ)," our empirically verifiable hypothesis.

Such is the general or abstract form of the Periodic or Moral Law. Let us now consider some of its concrete forms: abiotic, biotic, and cultural.

5. ABIOTIC COACTIONS

To map the chemical elements into this coordinate system (bottom part of Figure 5), we begin at the Axis of Atropy which bisects quadrant 2, and proceed counter-clockwise, representing the elements' increasing atomic numbers at 45° intervals.

The first Period has only two elements, hydrogen and helium. The second and third Periods begin on the Y_A axis and proceed around the coordinate system at 45° intervals, omitting Group VIII, but including Group 0.

It is important to understand why Period 1 consists just of Groups VII and 0, and can thus be regarded formally as beginning on the Y_A axis like the other Periods. (If one does not, this geometric mapping must seem arbitrary.) It happens to be a fact that two electrons (helium's number of electrons) complete the atom's smallest electron shell, the K shell. And it is shell-completion that makes atoms inert or atropic, the salient property of Group 0. It also happens that having one electron short of a completed electron shell (and hydrogen's one electron leaves it one short) makes atoms electro-negative, acidic, and extremely grabby, the salient properties of Group VII, the halogens.

Given these facts of nature, there is no other (physical) way for Period 1 to start. We are thus faced with two alternatives: Calling the halogens Group I, the inert elements Group II, and starting all Periods after the first with Group III; or doing what Mendeleev did intuitively: calling the first Group after the first Period Group I. I personally am glad that Mendeleev was obliged by the state of knowledge in his day to do the latter. Yet there is a dilemma, which some try to resolve in other ways.—Cassidy, for instance, places H in Group IV.[33] It is dilemmas such as this that make the world so fascinatingly and richly strange.

Beginning with Period 4 (the Period which has four Strata, four electron shells—K, L, M and N), another strange thing happens: The new outer shell, (N) starts to fill up. The next ten electrons, however, go into the inner (M) shell, where three of them start Group VIII, with three sub-Groups, a, b, and c. (This is shown empirically in Figure 5 top, geometrically in 5 bottom, and in Figure 11.) Then it goes on around to germanium, whose electron goes into the outer (N) shell, and so on to the double Period's completion with krypton, Group 0.—(We will discuss the Triple Group (VIII) when we come to the top Major Period, human cultures, where it concerns us most directly. Figure II-16b).—All Periods which circle the coordinate system twice in completing a new electron shell are called *double* Periods. They contain all nine theoretically possible Groups. The rest have either eight Groups or (the first Period) just two. It takes a big system (one belonging to a high Period) to develop and be able to survive Group VIII, the $(-, -)$ Group.[55]

Moral Tendencies Displayed by Atoms

The most cosmically important properties are those which, as Heisenberg points out, we regard as bad and good. Bad, he says, is the tendency to the confused and chaotic. Good is the tendency to increasing order or ectropy. Einstein probably referred to this when he said, "God is a tendency in the universe."

The chemical Group, relative to which these opposite tendencies can best be discerned, typically displays the (0, 0) coaction and is aptly numbered Group 0: the relatively inert or "noble" gases. (I would prefer to drop the latter name since, in my view, *Noblesse oblige*.) When mapped into the Periodic coordinate system, they fall on the circles of atropy, the 0 circle in quadrant 3. (See Figure 11 and Figure 5b.)[56]

The most highly cooperative, creative, and thus ectropic elements typically display the $(+, +)$ coaction. They are classed in Group IV, and are naturally mapped in the $(+, +)$ quadrant; namely, in quadrant 1 where Omega is located.

Carbon is the first and most famous Group IV element. "Organic molecules," says Harold G. Cassidy, "are the class of molecules that contain at least one carbon atom. There are more than a million different members of this class." And he goes on to discuss the brilliant electron-structure that makes carbon the most ectropic element.[33] Carbon atoms, Cassidy shows elsewhere, played and still play the key chemical role in biopoesis, the organization of life on our planet.[28]

The second Group IV element, silicon, plays the key role in the formation of life's habitat, the Earth. "Silicon plays an important part in the inorganic world, similar to that played by carbon in the organic world," says Linus Pauling. "Most of the rocks that constitute the earth's crust are composed of the silicate minerals of which silicon is the most important elementary constituent ... the *framework* minerals (hard minerals similar in their properties to

Generalization of Mendeleev's Periodic Table

quartz), the *layer minerals* (such as mica), and the *fibrous minerals* (such as asbestos)".[57]

And so forth through the existing Group IV elements and beyond them to the furthest ones envisaged for future creation or discovery. The "islands of stability," the theoretical regions in which the largest atoms are considered possible belong, Glenn Seaborg says, to Group IV.[21]

Turning to the chemically disintegrative, entropic elements, the most dramatic example is perhaps the halogens, Group VII, whose typical coaction is amensalism $(-, 0)$. These atoms, lacking one electron for completion of the outer shell or Stratum, attack many large molecules and destroy them in filling their need. "Fluorine ... is the most reactive of all the elements," says Linus Pauling. "Substances such as wood and rubber burst into flame when held in a stream of fluorine, and even asbestos ... reacts vigorously with it and becomes incandescent."[57] The next element in Group VII, chlorine, was the first poison gas of World War I. And so forth.

Some chemists question the classification of hydrogen in Group VII because of its creative (rather than destructive) role in so many life processes. The explanation, however, involves us in an Irish bull: If straight hydrogen (H) occurred abundantly in nature, nobody would ask this question for there would be no chemists. One of the things that make them and other life forms possible is the order in which the elements and molecules were formed: H atoms were the first to be formed in the expanding quasar shells, Figure 2. Their great reactivity or grabbiness made them combine into the first molecules, H_2. On one hand H_2 molecules behave not like halogen atoms but like Group IV elements. And on the other hand, the lightness and smallness of H atoms permits larger atoms such as carbon to transform grabber into grabbed, thus (to anticipate) changing potential amensalism $(-, 0)$ into symbiosis $(+, +)$, and potential entropy into ectropy. One function of unified science is to elucidate questions whose answers transcend the boundaries of one-field disciplines.

Space limitations permit us to discuss only one chemical Group displaying each of the primary moral tendencies: atropy, ectropy and entropy. The empirical reason why there are nine (and only nine) Groups has, to my knowledge, been discovered only implicitly, and thus unconsciously: It is implicit in Willard Gibbs' law that, given enough time and energy, a system will go through all of the states of which it is capable. In the expanding quasar shells there is enough time and energy for the whole atom population to ontogenate. (It did not evolve, as species do, but ontogenated as an individual organism does).[28] Gibbs' law may thus be called the Law of Ectropy, the counterpart of the second law of thermodynamics.

An explanation of Periodicity (with a capital P) is implicit in the Law of Ectropy: since there are only nine coactions, not more than nine Groups can constitute a Period. If a system continues to grow, some or all of the same Groups have to recur in each higher Period; another Stratum is thereby added to the one or more Strata already in existence, displaying the same Groups.—Whenever, on the other hand, the top Stratum's last remaining Group's entities disintegrate, the system declines to a lower Period. Grouping, Stratification and Periodicity are thus interdependent. And that, from bottom to top of the System-hierarchy.

Because of their strong and justified reaction against anthropomorphism, physical scientists have been reluctant to discuss, let alone to name the coactions displayed in the abiotic kingdoms.[58]

A geometrically coded vocabulary has, however, been developed for biotic systems, as shown below in Figure 12. These coaction terms may be extended thence to abiotic systems by readers who see their way clear to doing so.

6. BIOTIC COACTIONS

Proceeding visually from the abiotic (the lifeless) systems up the Systems-hierarchy (Figures 1 and 7) to living ones, the coordinate system remains constant, while the terms mapped into it become first those of biology, and then those of political theory. Historically, however, the coactions were discovered in biotic systems (Langlois' societies of fishes mentioned above), completed theoretically as in Figure 6, and then extrapolated "downward" to abiotic, and "upward" to cultural systems. This seems to fulfill predictions made in various ways by philosophers of science such as Warren Weaver,[4] and Suzanne Langer,[59] that synthesis would probably start with biology somewhere near the center and extend thence in both directions.

What made these extensions possible is the apparently universal validity of the relation which Niels Bohr called the *correspondence principle*.[60] He demonstrated wide and significant correspondences between the laws and relations which govern geoid systems (Major

Period 4), stated in terms of classical or Newtonian physics, and those which govern particles and atoms (Major Periods 1 and 2), stated in the very different terms of atomic physics. Bohr was forced to adduce this principle by the insuperable difficulties presented by the disparities between the Newtonian theories for geoid systems and those of Planck, Einstein, Heisenberg, Schrödinger, Pauli, Fermi and others for particle and atom systems.—And the data, reality, permit it. Bohr was thereby foreshadowing the present scientific revolution: One paradigm of unified science is the assumption that correspondence is universal, that it extends through the entire System-hierarchy. This belongs to a system of paradigms which reverses previous assumptions. (See Chapter V.) Thus it is the hallmark of what Kuhn has defined as a scientific revolution.[61]

Some major correspondences between the biotic and the atomic Groups have already been indicated. Biologists, however, have long had their own vocabulary, whose major terms are here mapped into the Periodic coordinate system.

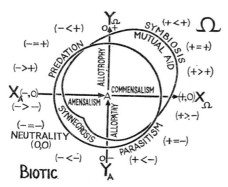

FIGURE II-12 Biotic Coactions Mapped into the Periodic Coordinate System. Since no terms for $(-, 0)$, $(0, -)$ and $(0, +)$ exist in the literature, they had to be coined. They have since been discovered in nature: plant, animal, and human.

The coaction vocabulary contained in biological glossaries and dictionaries is rather indefinite, displays several kinds of signal fouling and, when mapped in this coordinate system, displays the absence of three Groups. Namely, amensalism $(-, 0)$, allopathy $(0, -)$ and allotrophy $(0, +)$.[24] All three missing coactions have since been found in nature many times; and that, in all human Periods as well as in one plant Period and two animal Periods.

In order to present a biotic Periodic table, it is necessary, however, to show correspondence not only of the Groups, but also of Periods, Strata and Sub-Strata with those of lower Major Strata. Just as the atoms' Periods build up cumulatively, Stratum by Stratum, so also do those of the biotic and cultural Major Strata. Their relationship is diagrammed in Figure 13 by Harold G. Cassidy.

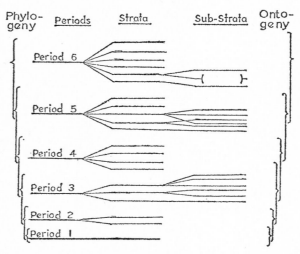

FIGURE II-13 The relationships of Sub-Strata to Strata, and of Strata to Periods in all Major Periods or Natural Empires. Abstracted by Harold G. Cassidy.

In all the natural kingdoms, the first Period has one Stratum; the second Period has two Strata; the third Period, three, and so forth. For all of them, the rule thus is that for each Period, the top Stratum's number is the number of the Period. The same holds true for the Sub-Strata: for every Stratum, the top Sub-Stratum's number is the number of the Stratum.

The Characteristic Numbers and their Assembly into Maps of the Web of Life

When Mendeleev classified the chemical elements, their number was well under a hundred; and it is not much over 100 to this day.[62] What we propose now is to classify *hundreds of thousands* of not just organisms, but of complex ecosystems; and to do so in a similar way.

How is this possible? It is possible if a method can be proposed, as will be done below, whereby scientists can team up to collect strategic data systematically and cumulatively, and feed them into modern computers. The computer can integrate these data periodically into a model of what ecologists call the web-of-life. It can simulate the operation of this web in such ways as to reveal the data's most strategic gaps and errors. Thereby it can direct the complex scientific effort in coherent, practically useful ways toward ever more accurate and complete models of the web of life on Earth.[63]

The object will, I hope, remain what it has been from the beginning of this undertaking: To help Man to gain, under God, control over his destiny. Toynbee has seen this sort of thing happen repeatedly, and calls it the Genesis of Civilization.

The first question which scientists will probably ask is this: What are the parameters, the variables, for this cumulative mapping of the web of life?

They are the variables which the Grand System-hierarchy's systems have in common; the ones in which, as Bohr would probably have put it, they correspond.[60] There are at least five such parameters. And they yield, when put together, the strategic construct which Leibniz apparently had in mind when he said, "We can assign to every object its determined Characteristic Number".[64]

Physical scientists have long been writing parts of such numbers for atoms. It is natural therefore that it should be a chemist who proposed the pattern here adopted: Harold G. Cassidy. The preceding diagram by him displays most of the pattern: Sub-stratum, Stratum, Period.

To these three parameters we add Mendeleev's Groups; but, naturally, couched in their general, geometric notation. And thereto, finally, we add the number of the classificand's Major Stratum or Periodic table.

Assembly of these five parameters of any given natural system into a conventional pattern results in its "determined Characteristic Number." These characteristic numbers can then be assembled into working models of portions of the web-of-life; and these can in due course be assembled into a world-web model.

The web of Characteristic Numbers lies one level higher than the Periodic Tables within the Systems-hierarchy of meta-theories or, as Quine puts it, background theories: The web of characteristic numbers contains all of the Periodic tables, plus one or more additional entities which have emerged, mutually modified.

The two parameters classified in Mendeleev's table are *Period* and *Group*. In adding the other three parameters to his Periodic table—and to the other Periodic Tables constructed in its image—we develop his table and Periodic Law further, fulfilling the prediction he made in 1889 before a joint meeting of the Royal Society and the Chemical Society: "It [the Periodic Law] needs not only new applications but also improvements, further development, and plenty of fresh energy. All this will surely come . . ."[20,65]

The general form proposed for all Characteristic Numbers is shown in Figure 14a; a concrete example of a web of Characteristic Numbers is then displayed in Figure 14b.

Practical implementation of Characteristic Numbers appears in the mapping of any portion of the web-of-life in any given space-time region. These relations have not been mappable in the past because of the absence of a systematic meta-language. This meta-language has now been developed, as demonstrated in Figure 14b. By itself, however, it would be practically useless: The web's enormous complexity and rapid dynamism far exceed Man's mental

STRATUM
(Ontogenic Ceiling)

SUB-STRATUM   MAJOR PERIOD   GROUP
(Current ontogenic stage)   *(Natural Empire)*   *(Characteristic Coaction)*

PERIOD
(Sum of Strata)

FIGURE II-14a   The Characteristic Number, mapping any process (except molecules and geoid systems) into the Periodic coordinate system. Designed by H. G. Cassidy and E. F. Haskell. Term coined by Leibniz.[64]

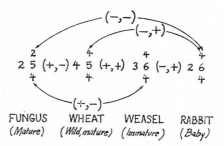

FUNGUS   WHEAT   WEASEL   RABBIT
(Mature)   *(Wild, mature)*   *(Immature)*   *(Baby)*

FIGURE II-14b   Sample of the Web of Life, mapped in terms of Characteristic Numbers and designed for eventual computer programming.

Generalization of Mendeleev's Periodic Table

capacity to grasp (diagnose), retain, and usefully modify it in real time. Modern computers, however, are capable of handling such problems. We need therefore have no anxiety as to the practical applicability of the web of characteristic numbers, no matter how complexly and rapidly it changes.—For modern Man, the impossible just takes a little longer. And this has been, and will be increasingly, an international modern undertaking.

Our little example maps a small part of a great world problem: the plant rusts.—These comprise some thousands of species of fungi, *basidiomycetes*, one or more of which live upon almost every species of seed plant, as well as some ferns. The rusts are probably, phylogenetically speaking, mosses which have degenerated through parasitisation of higher plants. Wherever they become dominant in a high (fourth Period) plant ecosystem, they destroy the highest plant Stratum (seed plants) and break the ecosystem down to plant Period 3; and when they become dominant in Period 3 ecosystems (whose highest Stratum is occupied by ferns), they break them down to plant Period 2, whose highest Stratum, 2, is mosses, the parasites' own highest ancestors. In whatever natural empire a lower Stratum destroys the controller and takes its place, this kind of breakdown follows cybernetically.

Our example is deliberately confined to a plant-animal ecosystem, Major Stratum 5. We are thus dealing with *wild* wheat. (Man, if he were included, would belong to the pre-agricultural human Period, 1.) In following this presentation, the reader is invited to refer to the framework of the animal Periodic table, Figure II-15; and to Figure I-6, diagramming plant Periods and Strata.

In the coaction-web drawn in Figure 14b, the coactions of four organisms are related to each other qualitatively as they occur at a given point in space-time. The coaction symbols +, 0 and − indicate positive, zero, and negative relationships. Numerical values can be ascertained and assigned to these symbols,[28] but are here omitted both because they have not yet been ascertained, and for the sake of initial simplicity.

The fungus' Characteristic Number is linked to that of its host, the wild wheat plant, by the symbol for parasitism (+, −) in Figure 12. This fungus belongs to the plant kingdom (5 at the center of its Characteristic Number); is situated in a seed-plant ecosystem (plant Period 4, written at the bottom); belongs to Stratum 2 of that Period (written at the top); and is in its second ontogenetic stage (Sub-stratum 2, written at the left).

Its host is a plant (5 in the center), belongs to the top plant Period (4, bottom), to its highest Stratum, seed plants (4, top), and is at its highest ontogenetic stage (4, left). This wheat plant is linked with the Characteristic Number of the weasel by (+, +), the symbol for symbiosis, for reasons which will appear directly.

The weasel is an animal (6 in the center); belongs to the fourth animal-ecosystem Period (4 at the bottom). (The Periodic Table of animal ecosystems is shown as Figure II-15.) The weasel belongs to its highest Stratum (4 at the top); and, while able to hunt, is not yet fully grown (3 at the left). Its Number is linked to that of the rabbit by the symbol for predation (−, +). (This is a standard symbol, Figure 12. Its form is correlated to its position in the coordinate system.)

The rabbit is an animal (6 at the center), in the fourth animal-ecosystem Period (4 at the bottom), the fourth Stratum (4 at the top); and is just old enough to get about (2 at the left).

The wheat-rabbit coaction is shown as predation (−, +), but could have been shown as parasitism (+, −) almost equally well. The essential relation is, that the wheat is damaged by rabbit (−), which is benefited by wheat (+); and this can be shown either way.[66] Since the weasel checks a damager of wheat, and since the wheat benefits the weasel's prey, the weasel-wheat coaction is (+, +). Conversely, since fungus and rabbit damage each other's food (wheat), their coaction is synnecrosis (−, −). Finally, the fungus-weasel coaction is estimated to be parasitism (+, −), since the weasel protects the wheat (the fungus' food), but the fungus damages the food of the weasel's prey, the rabbit.

This example shows the following: *Characteristic Numbers represent, not empirical species but ecological roles.* (This is the great advantage of the Periodic Table of chemical elements, whose entries consist of partial Characteristic Numbers.) This is what made it possible for Mendeleev and Seaborg to predict the discovery or invention of missing elements.

The greatest difference between atoms and organisms lies in the vastly greater number and variety of organismic species that can play any given ecological role, and in the changes of ecological roles which species can display. This makes it necessary to add the organism's traditional name to its Characteristic Number; and frequently to include additional information besides. For instance, while the rabbit-weasel coaction is predation (−, +) in the short run (synchronically), in the long run (diachronically), their coaction is

Generalization of Mendeleev's Periodic Table

symbiosis (+, +): rabbit-weasel population-explosions are reciprocally prevented. (The phased cyclic oscillation of prey-predator populations are typically cybernetic in character.)

The cardinal feature of this, and most other ecosystems, however, is the fact that each one has a *controller* (the wheat in this case) and a *work component* (the other organisms here represented). This system's relations are all determined by, and defined relative to, the wheat. In ecological terms, it is the ecosystem's *dominant* organism.—The clear distinction between controller and work component is basic to the diagnosis of an ecosystem's condition, and to the prognosis of its development. Should one of the lower organisms, say wheat rust, become dominant (become the controller), the ecosystem would break down to its lower Period. Conversely, where a high organism comes to dominate (control) a low ecosystem, the system transmutes up to the controller's Period.—In Chapter IV, this cybernetic principle will reappear in human cultures, resolving otherwise quite unresolvable political confusion, and consequent disputes.

Anyone familiar with computers can see at once how accurately this meta-language, when developed, should permit computer mapping and simulation of the web-of-life. This is a contribution obtained by geometrization of coaction vocabulary.

Students in inter-disciplinary courses which use this meta-language become quite expert at writing and interpreting the Characteristic Numbers of particles, atoms, plants, animals and men. They learn herewith to locate any of these natural systems, including particular persons at various points in their life-runs, within their own Periodic tables, and these within the Periodic co-ordinate system as a whole.

All parts of the Characteristic Number written in Arabic numerals belong to the left-hand part of the General Periodic Law, $R = f(\theta)$. They belong to the properties of natural systems R which are functions of coaction (θ). Stated another way, the Arabic numerals comprise the scalar aspect of the radius vector, the Roman numerals (or their geometric counterparts) represent the coactions of which these properties are functions.

The Periodic Table of Animal Ecosystems

We turn now to the first step in the cumulative assembly of Characteristic Numbers into the framework of, for instance, a biotic Periodic table. In Figure II-15, an entire biotic Periodic table is

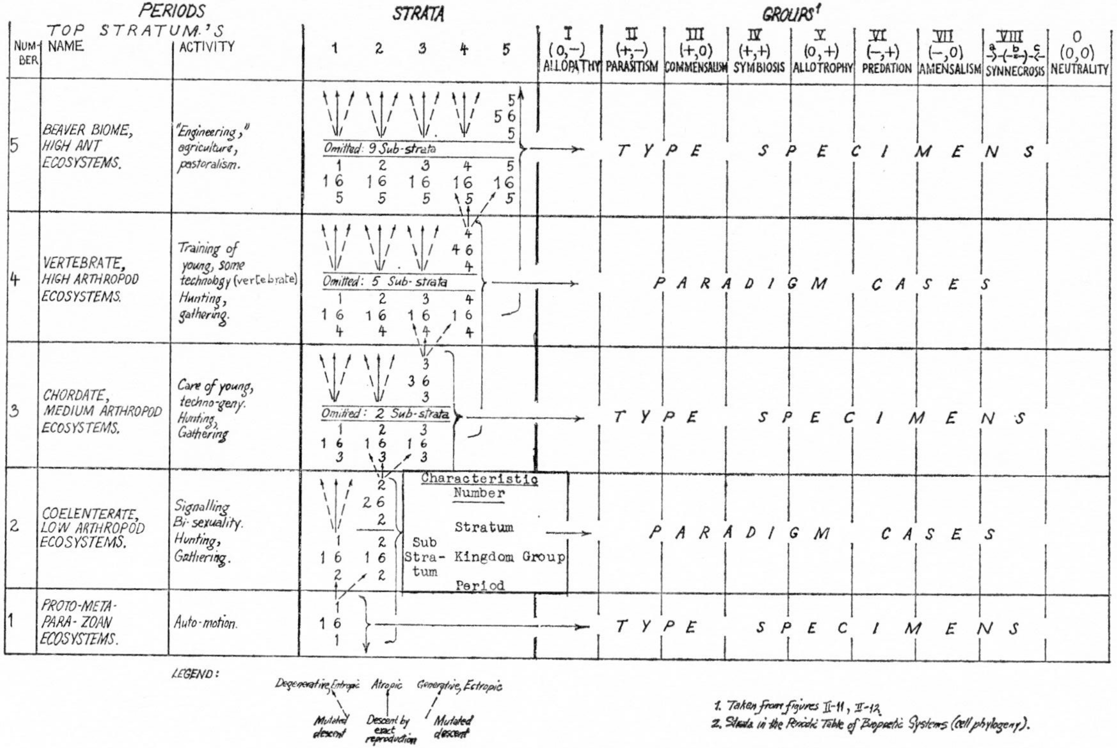

FIGURE II-15 Framework of the Periodic Table of Animal Ecosystems, Major Stratum 6, in terms of Characteristic Numbers.

Generalization of Mendeleev's Periodic Table

displayed in token fashion: the Periodic table of animal ecosystems. (The framework of the Periodic table of human ecosystems will be shown in Chapter IV.) These frameworks consist essentially of arrays of what biologists call *type specimens*, and Quine calls *paradigm cases*. In all these frameworks, however, the specimens are systems; and usually ecosystems. For it is never the organism alone that evolves; it is always, and only, the ecosystem: the habitat-organism system. Collecting these specimens will require appropriate equipment and techniques; and storing the data will involve pictures, films and tapes, besides paper-filing cabinets. The framework of the Periodic table of animal ecosystems is seen opposite.

Across the top right half of this framework, the nine Groups are listed and numbered with Roman numerals in their traditional order; and the Periods proceed upward at the left from the bottom to the top.

Each Period has a four-part description, ordered in four columns: The column at the far left lists the Periods' numbers. Next to it is listed the name of the animal Phylum which occupies the top Stratum of the Period in question. In the third column is listed in each case the new, emerged process which these animals display, and which distinguishes them from the previous Period's highest animals. And in the right-hand column, are stated the four scalar (Arabic-numeral) parts of the Characteristic Number of the particular organism in question. This summarizes the first three columns.

Period 1 is thus labeled *Proto, Meta, Para-zoan ecosystem*; its characteristic process is *Auto-motion*; and the scalar part of its characteristic number, in the fourth column, is $\begin{smallmatrix} 1 \\ 1 \ 6 \\ 1 \end{smallmatrix}$. Its nine possible Groups then follow across the right-hand side of the table, each to be labeled with the reference to a list of the animals, plants and men which typically display the coaction in question with the organism of reference.

Each higher Period, as we proceed up the table, contains the preceding Periods (usually in modified forms) as its lower Strata. Among these modifications are each lower Period's mutated forms, thousands of which are represented in token or framework fashion by the dashed slanting arrows in the fourth column. (The extinguished lines-of-descent do not appear.) Leftward-slanting dashed arrows represent degenerative (entropic) evolution; rightward dashed arrows represent creative, ectropic evolution; and solid vertical arrows indicate atropy: zero or negligible evolution.

All characteristic numbers in this table (except the upper right-hand one in each Period after the first) denote zygotes, the lowest Sub-stratum marked 1 at the left. The upper right-hand one, however, represents the mature adult. (The Sub-stratum attained is here equal to the Stratum ceiling, the number at the top.) Where necessary, as in the case of the stages of insect metamorphosis, Sub-strata can and should not only be entered numerically, but should be keyed to their traditional scientific names as well.

Since this ecological classification is based upon, and includes, all organisms listed in the taxonomic series, it makes use of all the achievements of genetics and evolution theory. (It thus includes all of what is traditionally called Natural Law.) Yet it decreases the penalties incurred by the taxonomic series' omission of habitats and of the moral or tropic tendencies.[67]

The Periodic tables of plant and animal ecosystems include all of pathology and immunology related to *parasitism* (Group II), all of *predation* and the defenses against it (Group VI), all of co-operation, mutual help and synergism subsumed under *symbiosis* (Group IV); and so forth through the nine theoretically possible and empirically recognized coactions and their ranges of gradations. It thus includes a considerable portion of what is traditionally called the Moral Law, and states it in scientific terms. In consequence, these tables can be made to present part of a framework for the coherent, systematic study of the web-of-life's moral or tropic tendencies relative to the highest member of its System-hierarchy, human cultures. (See the Appendix at end of book.)

7. CULTURAL COACTIONS

We turn now to what Mendeleev called "the relations of man—social and political," which he predicted could be included in "the general reign of order in nature, and in the entire universe as a whole."[20]

A few years before Langlois discovered that the properties of bass societies are functions of the coactions between the Majority and the Minority, Arnold Toynbee had, also unconsciously, discovered a similar version of the Periodic Law in more than twenty human civilizations.[19] He had found that during the periods when these two components had been symbiotic and helped each other, (+, +), their societies' major properties were these: high productivity, cohesiveness, and good health, especially mental health. He

Generalization of Mendeleev's Periodic Table 63

called these periods *the genesis of civilization*, and their Minorities *creative*. But when predation developed $(-, +)$, so that the Minority benefited greatly at the expense of the Majority, these properties sharply changed: mental health declined (*schizm of the soul*), social cohesiveness went into its opposite (*schizm of the body politic*), and productivity plummeted: on one hand a large part of it became armaments; and on the other, real goods were destroyed by these armaments. These periods he called the *breakdown of civilization*, and their Minorities *dominant*. At this point each of these twenty-odd civilizations went into what Toynbee called the *rhythm of disintegration*: it oscillated back and forth between predation $(-, +)$ and parasitism $(+, -)$, finally reaching the climax of committing suicide by mutual destruction, synnecrosis $(-, -)$. Toynbee called this *disintegration of civilization* and showed that it regularly results in the transmutation of the civilization's breakdown-products into what we now call a lower human Period.

Toynbee discovered these opposite tendencies in Literate societies, which comprise human Period 5, as will be shown in Chapter IV. They have also been observed in a Lower Agricultural society (Period 2) by Gregory Bateson, and in Industrial societies, (Period 6) by many social scientists.

In his book on the Iatmul of New Guinea,[68] Bateson showed that primitive societies split or fission in two opposite ways which he calls *horizontal schizmogenesis* and *vertical schizmogenesis:* When they split horizontally, between the Majority or lower Strata and the Minority or controlling Strata (which usually involves negative coactions between the two), the society disintegrates. But when they split vertically, so that a large part of the Majority and Minority together leave the village (which usually involves positive coactions), both the old and the new village remain viable societies. (This peaceful, vertical schizmogenesis occurs routinely in many high animal societies such as those of bees. So also in Literate human societies such as the one which formed Australia, New Zealand, and Britain's American Colonies.)

A notable case of vertical splitting which has been reported for an Industrial society (Period 6) occurred in modern Switzerland.[69] A tiny group of commercial, industrial, political and social leaders (part of the Minority), together with a sufficiently large group of the broad public, the Majority, formed a Federation of Co-operatives (Migros), a political party (the Landring of the Independents), and an educational and research institute (the Gottlieb

Duttweiler Institute). In a series of brilliant, tenacious campaigns that began in 1925 and continue today as strongly as ever, they have forced Switzerland's Dominant or Caste Minority to lower monopolistic prices, thus changing predation $(-, +) \rightarrow$ symbiosis $(+, +)$; forced truth about monopoly prices, profits, wage demands, strikes, lockouts, and so forth into the press (often excluded in the name of freedom of the press); forced discussions of the same into Parliament (often excluded in the name of free speech); and turned repressive law suits into public forums. (Migros now has the largest press in Switzerland. Persistence of the vertical front in that country is therefore very probable.)

The best study, for the United States, of Bateson's classical concept "vertical and horizontal schizmogenesis" is Baltzell's famous book, "*The Protestant Establishment*—Aristocracy and Caste in America."[70] But perhaps the most *famous* cases of the prevalence of the vertical over the horizontal front in the English speaking world are the Magna Carta and Cromwell's Protectorate. In neither case was the Dominant Minority destroyed or liquidated. Instead, it was forced by a Creative Minority to change its predation $(-, +)$ into a considerable degree of cooperation $(+, +)$. This strategic and creative maneuver resulted both times in a Genesis of Civilization.

The most famous cases of the prevalence or "victory" of the horizontal front, on the other hand, are the French and Russian revolutions. In both cases, a Dominant Minority was *destroyed*. The society was thereby decapitated to such a degree that it has not recovered. It resulted, in the case of France, in perpetual oscillation between republics and empires; in that of Russia, in perpetual and spreading dictatorship and terror.[71] It thus appears that human cultures, emerging at the apex of the System-hierarchy, obey the General Periodic Law as clearly as do their preceding and accompanying biotic and abiotic systems.

Historical forces and accidents, however, have resulted in systematic though unconscious errors of theory which transform what would otherwise be intellectual schools of political science and sociology, comparable to those in physical and biological sciences, into emotional ideologies. On one hand the tradition of the vertical front occurred as the growth of folk custom in the tradition, say, of British and American Common Law. On the other hand the tradition of the horizontal front developed as a theoretical system in the tradition, say, of French rationalism and German philosophies. These two political traditions are thus not only emotionally and temperamentally diverse; they are, in addition, couched in the two diverse and here-to-fore incompatible modes of thought: The first is carried on by way of what James Conant calls the *inductive empirical* mode of thought; the second by way of the *deductive theoretical mode*.[72] (See Figure V-I.) Both of these temperaments and modes of thought, moreover, occur not only in every industrial nation, but in each Stratum, and Sub-Stratum or age-grade of each industrial and even most sub-industrial nations.

Formulation of the strategic problem of the psycho-socio-political sciences, and resolution of their ever growing crisis and impasse, can occur only in terms of a discipline which defines and distinguishes these cultural categories, beginning with their simpler counterparts in the abiotic and biotic sciences; of a science, moreover, which has developed a method for cleansing these sciences' vocabularies of communication fouling: namely, the geometric definition of concepts and terms in Unified Science. The coaction compass can be boxed in the languages of the anthropo-socio-political as well as in those of other sciences.

The basically diverse historic meanings of the paradigmatic framework of political thought, *Left* and *Right*, were clearly displayed by the seating arrangements of the British House of Commons on one hand and of the French *Etats Généraux* in the year 1789 on the other. In the House of Commons (as in the U.S. Congress) the seating did not display social stratification; but it did in the Etats Généraux. What it displayed in the British case was (and remains) the following: on the Right, adherence to the Government at the time in question; on the Left, the Government's Loyal Opposition. Since all social Strata are represented on both sides of the Speaker and the aisle perpendicular to him, this aisle represents and demarcates a case of the *vertical* front: that Left and Right remain predominantly class-cooperative is guaranteed by the institution of universal suffrage by secret ballot. Members of Parliament *ipso facto* belong to the Minority. But they are elected freely by, and represent, the Majority. They cannot otherwise sit either on the Left or Right.

The composition of the Etats Généraux of 1789 was basically different: There *l'aristocracie*, the nation's Minority or controller, sat in splendor at the Right (of the rostrum), the *Jacquerie* (or Proletariat) in squalor on its Left, and the *bourgeoisie* between them, in the Center. The latter two Strata *together* represented the system's Majority or work component, with the much smaller Bourgeoisie ready to seize the control.

The coaction was different too: These basic components of the culture, Majority and Minority, had long been *alienated* from each other: *L'aristocracie* was a Dominant Minority, the Jacquerie and bourgeoisie an alienated *proletariat* in Toynbee's meaning of being in, but not of, the society. At this point in history they became enraged at each other. Their negative coaction approached the limit of ferocity. *And that negative coaction was fixated as the paradigm of political theory, and became the dominant value premise of large regions of the world.*[73] The King and all aristocrats who could be found were imprisoned and executed by savage mobs and tribunals. Any profession of *loyal* opposition became grounds for arrest. All possibility of a vertical, class-cooperative front was consciously destroyed, and the Proletariat rampaged destructively through the streets.

The manner in which order was restored reinforced the paradigm of class-conflict: It was not the introduction of a better (less predatory, more symbiotic) coaction, as in Switzerland, but artillery turned on the mob by a young officer, Napoleon Bonaparte. He represented, and restored control to, the Dominant Minority in another form, the Bourgeoisie; a form which presently evolved with equal savagery into what Talmon calls *Totalitarian Democracy*,[74] controlled by what Milovan Djilas calls *The New Class*.[75,76]

Had the *vertical* front of class *cooperation* been generalized into a theory about "all of recorded history," it would have been incorrect, for no one coaction ever has excluded all the others. Yet it is positive coactions that *predominate* on our planet. They have predominated in all natural Kingdoms for billions of years, as shown by the fact that evolution has been mainly *upward*. And all the great religions unite in affirming it. The largely class-cooperative British and Americans however, as Conant has pointed out, tend to be weak in the deductive-theoretical mode of thought.[72] Their Creative Minority therefore failed to produce a coherent, let alone a compelling, socio-political theory demonstrating the value of positive coactions. Baltzell shows this most clearly in his chapter on "The Intellectual Counterattack on Caste: The Social Gospel, Reform and the New Social Science."[70]

The English-speaking Dominant Minority, it is true, developed no more coherent or compelling theory of horizontal splitting, as Baltzell shows in his chapter on "The Ideological Defense of Caste."[70] But their angry colleagues in Continental Europe, being traditionally prone to the deductive-theoretical mode of thought,[72] picked out and exalted local experiences of horizontal, conflictive schizmogenesis

Generalization of Mendeleev's Periodic Table

into what they consider to be a universal axiom: "All of recorded history is the history of class conflict".[77,78]

This paradigm they reinforced in both theory and practice: They deduced theoretically that the only way to end exploitation $(-, +)$ is to destroy the controller, the Minority, which they have done throughout over a third of the world. And they deduced that the only way to generate symbiosis $(+, +)$ is to develop a "control-less" ("class-less") society. This goal being objectively impossible—and they *themselves* having, of course, assumed control, as Djilas has shown in *The New Class*[75]—this Dominant Minority enacted its goal *subjectively:* it called itself "The Vanguard of the Proletariat," where "proletariat" is defined as the first three social strata. Being convinced that the only class relation possible to Mankind is conflict (and being involved in it themselves), they fear and punish their identification as a class, and declare their misinterpretation of history to be "irrevisible."

Forced by their clearly incorrect theory on one hand to imprison or execute those who show its inconsistencies, and on the other hand to substitute in place of objective reality and intellectual force a vast apparatus of radios, jammers and agitation, the horizontal front has filled the upholders of vertical fronts around the world—which may well be the silent majority—with deep anxiety. Being unable to defend itself theoretically for the reason just given, it tries to do so militarily, involving itself too in contradiction. Both fronts are thus involved in ever more fearfully dangerous, protracted and self-intensifying politico-military conflicts.[79] And today, in the Seventies, they are as clearly bringing on World War III as in the Thirties they were bringing on World War II. But World War III would probably fulfill the old and persistent predictions of Apocalypse: It would destroy the Earth's System-hierarchy all the way down to radioactive, lifeless rocks and oil-slicked water which looks not altogether unlike what these ancient predictions have described as "dead men's blood".[80]

This is, of course, a total contradiction of the appealing course of events predicted by the horizontal front's glib theoreticians. But these conflict-prone Left and Right theoreticians are as helpless to change their course as are the less articulate upholders of vertical front; its "silent Majority" and equally silent Minority. For what is true of scientists applies even more strongly to ideologists: "What scientists never do when confronted by even severe and prolonged anomalies", says Thomas Kuhn, "[is to] renounce the paradigm that has led them into crisis . . . Once it has achieved the status of paradigm, a scientific theory is declared invalid only if an alternative candidate is available to take its place."[61]

As chairman of a 1936 delegation observing fascism, communism and democracy in ten countries, I was forced to realize that the conflict-dominated theories of the Extreme Left and Extreme Right—of Communists and fascists—were bringing on World War II. And it was perfectly clear that no matter who won that war, these theories would, *in the absence of a compelling alternative*, bring on World War III, as they are clearly doing now. It was this conviction which determined me to undertake unification of the sciences. For that is the only effective alternative to ideologies I could envisage.

I had no conscious idea that the cardioid of values would emerge in this sum of zero-valued parts. But the value-cardioid proves to inhere in Unified Science's very corner stone, the Periodic Law. Herewith it challenges both the zero value-premise of the one-field scientists and the negative value-premise of the ideologist; and it supports the positive value-premise of what Toynbee has called the seven great religions. Herefrom it follows, as the day the night, that the great schism caused by the rise of the sciences has come full circle, and that the mental and spiritual crisis of industrial civilization is finally capable of coming to an end.

The schizm began, as Authur Koestler shows, when the empirical scientists and theoretical theologians agreed to go separate ways: the scientists agreed to confine themselves to *res extensa*, the theologians to *res cogitans*, splitting the West into the Two Cultures.[81] This crisis developed into chaos as the sciences on one side grew apart from each other, and the ideologies and religions on the other side parted from each other too.

The scientists rationalized their separations by assuming that abiotic, biotic and cultural processes differ basically from each other. And humanists supported this assumption by declaring that Moral Law and values are confined to Man, and are not shared by animals, plants, and non-living entities. The groundless axiom of each of these sub-cultures supported that of the other in forcing industrial civilization's breakdown, and preparing conditions for the rise of the negatively biased ideologies.[82]

Unified Science's paradigm is opposite to those of both of our traditional sub-cultures: The structures of all systems—abiotic, biotic, and cultural—are assumed to be isomorphic unless proved otherwise, which has certainly not been done. The laws of the simplest systems,

Generalization of Mendeleev's Periodic Table 69

those belonging to Major Stratum 1 and 2, are assumed to be teleonomic; to be so structured that their *telos* or goal—which is the highest Major Stratum, human culture—is implicit in their System-hierarchic combinations. The Periodic Law of chemical elements is shown to be a special case of a general law which rules throughout; namely, the Moral Law.

In 1889 Dimitri I. Mendeleev foreshadowed all the paradigms of Uunified Science in his above mentioned address before the Royal Society of London: "There are in the world two things which never cease to call for the admiration and reverence of man," he said, quoting Kant: "the moral law within ourselves, and the stellar sky above us. . . . But we must add a third subject, the nature of the elementary individuals which we discover everywhere around us. . . . In the atoms we see . . . the submission of their seeming freedom to the general harmony of nature."[20]

This law—extending from Man through the hierarchic sky down to the atoms—affirms that *the properties of systems are functions of their coactions;* that "As ye sow, so shall ye reap." This contradicts the paradigm of the Cultural Relativists who affirm that cultures with diametrically opposite value-premises are equally valid.[83] In the same way it contradicts the Existentialists' paradigm that there are no objective values; that it is therefore indifferent what values the individual may choose to follow.[84]

When I say that the world has come full circle I mean that we have come to recognize that *Moral and Natural Law are the same thing.* A person is free to defy them both subjectively. Objectively, however, he is free to select the degree of grace with which he submits to the general harmony of nature. No lower member of the Systems-hierarchy has such freedom.

8. THE COORDINATE SYSTEM OF POLITICAL SCIENCE

The so-called political spectrum has served for nearly two centuries as political science's frame of reference. Yet it is flagrantly incorrect.

EXTREME	FAR	MODERATE	RIGHT	ROSTRUM CENTRE	LEFT	MODERATE	FAR	EXTREME
Right	Right	Right	Center		Left	Left	Left	Left

FIGURE II-16a   The Political Spectrum; 18th century construct which time has shown to be radically misleading.

70 Full Circle

Its criticism and correction will be carried out in two complementary ways, theoretical and empirical.

The deductive-theoretical mode of thought permits short and decisive treatment. The situation can be formulated geometrically, thus: The traditional political parties represent the principal resultants of two variables, X and Y. These cannot be deployed correctly on a straight line because the limiting resultants (at the ends of the line) are actually adjacent. Their correct deployment requires the line to be curved so that its ends adjoin.

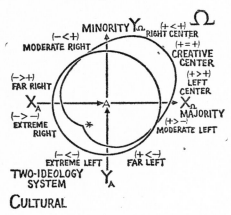

FIGURE II-16b The Coordinate System of Political Science: obtained by mapping the Political Spectrum into the Periodic coordinate system[71].

The concept, *political spectrum*, resulted from a false analogy between the array of political coactions and the physicists' color spectrum. The color spectrum deploys a single variable, wave length. This extends from very short at one end of the straight line to very long at the other. This deployment is correct: Where the variable is arrayed perpendicularly to the straight line, the interval between shortest and longest variable corresponds to the interval between the straight line's ends, whatever its length.

Because this false analogy does not conflict with the system of incorrect paradigms underlying the array of fragmented sciences (see Chapter V), the analogy was accepted in the face of dramatic empirical contradictions to be set forth next.—We turn now to the inductive-empirical mode of thought.

*The locus of the Sub-Group VIII b (— = —), whose political name is *Nihilism*.[90]

Generalization of Mendeleev's Periodic Table

Empirical Criticism and Correction of the So-called Political "Spectrum"

For thousands of years, Toynbee has shown, the seven great religions have displayed positive value-biases. That is to say, they have advocated and stressed mutual help and class cooperation in various ways, degrees, and idioms.[85] With the rise of modern science since the 15th century (studying, as it had to, *parts* of systems), values were subjectively confined to the humanistic and literary subculture. However, what the rising scientific sub-culture actually (objectively) adopted was a value-bias. Namely, the *zero* value-bias. It claimed, subjectively, to have *nothing* to do with values. And since its various specialists dealt largely in isolated *parts* of systems—e.g. in habitat-less plants, animals, and people, in which the Periodic Law is *not* discernible (and since it did not recognize the moral nature of that law in atoms, where it *is* discernible, as Mendeleev intuited)—it confused its zero value-bias with *no* values, and banned discussions of morals from its leading societies.

Then, in the 18th and 19th centuries, emerged the *negatively* biased misinterpretations of history, advocating class conflict on the Far and Extreme Left, race and national conflict on the Far and Extreme Right.

Traditional scientists (who study just system-*components*), having renounced values, are helpless to interfere with ideologists in any effective way. Traditional men of religion—speaking, as Bishop Robinson affirms, in the no longer effective language of pre-industrial civilizations—have been engaged in mere rearguard actions for over a century.[86,87] And so, as the conflict-spreading propaganda apparatus penetrates the world's mass media—its films, television, radio, and press—the traditional spokesmen and their followers, the great majority, fall silent. For in a culture whose dominant value-premise is becoming negative, as it is now in ours, the people with positive value biases become deviant; and deviants tend to become silent, even when they are the majority.

What Unified Science now asks mainly of scientists (who are the best equipped to get the thrust of this question) is the following: Can we accept a frame-of-reference—a coordinate system such as, for instance, the so-called political spectrum—without considering the way it has been formed?

Consider the case of physical scientists before Einstein's theory of Relativity emerged and corrected their only locally correct Newtonian frame of reference. Would it be realistic to consider social scientists immune to similarly incorrect micro-centric points of view? I quote from a paper I presented at the Second International Congress for the Philosophy of Science:

> Einstein has shown that in physics, ideas of local physical phenomena are generalized into theories of the universe. (A perpetually rotating room or free-falling elevator, he shows, would give rise in its inhabitants to specially biased kinds of physics.) Similarly, autocratic and predatory cultures on the one hand, democratic and symbiotic cultures on the other, have given rise in their inhabitants to specially biased kinds of political philosophy. These intellectual biases we call logo-centrisms . . .
>
> The same principle has been shown . . . to operate on feelings and emotions. Societies with emotional "climates" of overwhelming fear and hate produce, in their inhabitants, philosophies of universal conflict and danger; societies with "climates" of friendly cooperation produce in their inhabitants equally biased philosophies of universal friendship. These emotional biases we call pathocentrism.
>
> The opponents within each political system share the same logo- and patho-centrism. Hence the relative ease with which they exchange roles, and the difficulty with which they change systems . . . The bearers of each of the patho- and logo-centrisms tend to attribute their own views and motives to the bearers of the other . . . in spite of contrary empirical evidence . . . This illusion is systematically utilised by the ideologists, to prepare their prospective victims for involvement in mental infection and physical conquest.
>
> The greater the proportion of conflict and falsehood in a political philosophy, the greater the isolation necessary for its continuation. As, in Einstein's hypothetical falling elevator or rotating room, opaque walls are essential to the maintenance of the inhabitants' special kinds of physics, so in the Two-Ideology system an "Iron Curtain" is necessary to maintain the inhabitants' mis-interpretation of the world and of history. The "Curtain" exists in fact.
>
> It follows that, with the development of geometric definition, mutual confrontation of the . . . systems—or their basic natures, structures, origins, functioning, and consequences—must tend to counteract falsehood and other negative relationships . . . Geometrization will therefore help the creative individuals among the

inhabitants of these local habitats . . . to evaluate them realistically, and hence to transform them . . ."[71]

Example:

Gilbert Allardyce has compiled descriptions of a single political phenomenon, *fascism*, as seen from three frames-of-reference: the communist Left, the humanist-socialist Center, the conservative Right. (See "Fascism as the End of Liberal Society" both as treated by him in his preface and by the authors in the text of his book *The Place of Fascism in European History*.[88]) Several other interpretations of fascism are included in his book, most of which fall into various categories of Unified Science. (But they necessitate more detailed analysis than can be included in the present volume.)

The point which needs to be made here is that Unified Science uses the same method to reconcile the discrepancies due to diverse frames of reference which relativity physics has used so successfully: systematic transformations made possible by the emergence of an invariant background-theory and its geometric background language. (See Harold Cassidy's assessment of Unified Science in the opening chapter, made from the viewpoint of a physical scientist.)

The profound and pernicious incorrectness of the one-dimensional "political spectrum", Figure II-16a, has been sensed by political scientists for nearly fifty years: "Some scholars," says Allardyce, "had already connected fascism and bolshevism in the 1920s, almost from the moment that the Blackshirts appeared on the Italian scene. Being for the most part men of liberal and democratic opinion, they associated Mussolini and the Bolsheviks with a common assault upon free institutions and open societies. It appeared to some of them that the terms 'Left' and 'Right'—descriptions which had never been very satisfactory anyway—no longer defined political reality, but rather seemed to obscure it." p. 13.[88]

This was the clear recognition of profound anomaly which typically characterizes the outbreak of crisis in any science. And here as in other sciences, it is a coherent scientific theory, based upon an incompatible paradigm, which resolves the basic anomaly and solves a number of troublesome subsidiary problems, which marks the outbreak of a scientific revolution.[61] Such is the case at present, as I will now try to demonstrate:

What obscures political reality is not the terms *Left* and *Right*, which correspond to fundamental cultural structures, but the failure of that 18th century frame-of-reference to group together the *symbiotic* (Center) Left and Right, to group together the *synnecrotic* (Extreme) Left and Right, and to separate these fundamentally *diverse* pairs of phenomena *from each other*. This greatest of all moral distinctions, and the intermediate distinctions between these limits, appear when the traditional simplistic, amoral 18th century political frame-of-reference, Figure II-16a, is mapped into the Periodic coordinate system, the frame-of-reference of the Moral Law, as shown in Figures II-16a and b.

The political scientists' schools array themselves just as neatly as schools formed during the crises of other sciences: Those ideologies which Digby Baltzell groups under "The Ideological Defense of Caste" fall on the Right of Center; those which he classes under "The Intellectual Counterattack on Caste" fall on the Left of Center. The Right Center schools stress evidence connecting social status to genetically determined, hereditary traits, and minimize evidence connecting social status to environmentally determined factors. The Left Center schools, on the contrary, minimize evidence connecting abilities and social position to genetically determined, hereditary traits; and emphasize evidence relating environmental factors to social position, low and high. Those Leftist ideologies holding positions up to and including the Moderate Left are egalitarian and culturally relativistic (that is, anthropologically egalitarian); those up to and including the Moderate Right believe in what Thomas Jefferson called *Natural Aristocracy;* namely, that people born with outstanding talent and virtue (like himself) should, and usually do, occupy controlling social positions.[89]

Motivation:

Egalitarianism and environmentalism subjectively raise lower Strata and lower Periods relative to higher Strata and Periods, grinding their incumbents' axe, and making Left theorists their spokesmen and leaders. On the other hand, minimization of environmental effects, and exaggeration of hereditary factors increase the self-assurance (subjective power) of those occupying controlling positions, and decrease that of those occupying lower Strata. This grinds the top Strata's axe and makes Right theorists their spokesmen and intellectual leaders.[70]

What these schools *share*, however, is far more important than that which separates them. Namely, the positive value-bias summed up in the commandment, Do unto others as you would have them do unto you.[47] All the political schools of the Center listen to each other's view and evidence respectfully (though sometimes reluctantly), credit each other's sincerity, and trust each others' loyalty to the principles of free speech, press, assembly and the ballot. Symbiotic freedom is the expression of love, of positive coaction. That makes inequality, which is fundamental throughout the universe, *creative*. Hence the term *Creative* Center. At the Center, Left and Right are convinced of their common positive value-bias and thus of their loyalty to each other in opposition. These are the criteria of the most important political system.

As we proceed toward the Left and Right extremes of this frame of reference, however, we note ever increasing vehemence and violence, ever decreasing interest in and respect for evidence, and ever decreasing mutual respect and trust. (See, for example, R. Palme Dutt's chapter in Gilbert Allardyce's book, just mentioned.)

And a certain point, called "loss of legality", a quantum change occurs, sudden and immense: suddenly the man, whether of Left or Right, finds himself not with just a vested interest, but with his very freedom and survival themselves, or those of his enemy, (Lenin's "Who-whom", dependent upon his own political control. Under these circumstances, the life, liberty and happiness of both the Left Extremist and the Right Extremist leaders depend upon the other controller's destruction, and upon the incorporation of his opponents' followers (their work components) into his own system.

This is, as Americans characteristically put it, an entirely *different ball game*: the Extreme Left and Extreme Right are playing for keeps, with no holds barred. They understand each other, and they fear and respect each others' ruthlessness and cunning. They are, of course, utterly contemptuous of the trustfulness, tolerance, and relative truthfulness of all Center parties. (They call them "naive, sentimental, vacillating, reformist, hypocritical," and so forth.) They enter transient "united front" alliances with Center parties against each other. They also enter temporary alliances with each other against Center parties, each closely watching for the opportunity to stab his well understood, highly perfidious "ally" in the back.

The almost certain outcome is that of the Two Ideology system's most famous case, the Stalin-Hitler pact: mutual devastation or, as biologists put it, synnecrosis $(-, -)$. In a great historians terms, this is called *disintegration* of civilization.[19,90] Conversely, the probable outcome of continuous symbiosis of the Creative Center—of their vertical front as developed, for instance, in Switzerland[69] or in Benjamin Franklin's Pennsylvania[91]—is dynamic peace, prosperity, and *genesis* of civilization.

When the one dimensional political frame of reference is mapped into the Periodic coordinate system, as in Figure 16b, the similarities and differences which have long bewildered so many political scientists, and fouled up their communication, are fully and decisively cleared up.

This mapping shows that the difference between the Creative Center and the Two-Ideology system is very much greater than the difference between Extreme Left and Extreme Right. (The one-dimensional "spectrum" indicates precisely the opposite, seriously confusing political thought.) According to Allardyce,

> To interpret the fascists as a right-wing phenomenon, to seat them beside the monarchists and reactionaries [of the Center] in parliaments, and to consider them a radical extension of die-hard conservatism [rather than as qualitatively different from it] was to be blind to the remarkable similarities between Mussolini and his communist 'enemies.' [The semi-quotes symptomatize the anomaly in these thinkers' concept.] With the rise of fascism, it appeared, the heat and passion of both ends of the political spectrum [Figure 16a] had resulted in a fusion of political extremes—the ends had met, completing the circle of political beliefs [Figure 16b] . . .
>
> From these beginnings an inquiry has gone forward into the underlying unity of the radical movements. The result has been a continuing literature on the connections between the two political poles, with social scientists seeking the common roots of left and right extremism, and psychologists particularly involved in working out theoretical models of what has been called the 'authoritarian personality'.[92] But it was the dual experience of Hitler and Stalin, much more than the earlier confrontation between fascism and communism in Italy, which gave these studies their real impetus and significance. Indeed the sight of the dictatorships in Berlin [today, East Berlin] and Moscow evolving side by side has stimulated some of the most original political thought of our time. It has provided our mental imagery of the modern political state and revealed the existing possibilities for total power over men.[93] The two regimes opened our vision into the world of 1984, and returned the concepts of mystery and evil to political theory[88] pp. 13–14.

Generalization of Mendeleev's Periodic Table

That, however, is but a preliminary approach to the full circle. Full circle itself is reached when the Moral Law's concepts of *good* as well as of *evil*, its new intermediate concept of *atropy* between them and, more particularly, its concepts of the nine Groups and their many quantitative variations—when these are ordered geometrically in the same terms as those of the physical and biological sciences, Figure 16b. Being part of a scientific revolution, it has the power to change the course of history away from the formerly probable goal described as 1984 and toward genesis of the Higher Industrial Period.[91]

If, now, you will look once again at the so-called "political spectrum," Figure 16a, its anomalies will be unmistakable: Using a ruler or a compass, compare the distance between Right Center and Extreme Right, or Left Center and Extreme Left, with the distance between Extreme Right and Extreme Left. The latter is more than twice as great as the former distances, when precisely the opposite is the case in reality. Is it not clear that any science—abiotic, biotic, or cultural—must be in crisis to the degree in which its frame-of-reference contradicts and misrepresents its data, and that the crisis of political theory and political practice has, in fact, been profound and desperate to just that vast degree?

Until the rise of Unified Science, the one-field sciences' background theory, it has been impossible for all but the most astute and imaginative Centrists and Extremists to understand each other or credit each other's actual existence: each had developed strong *logo-centrism* and *patho-centrism*, and there was no concept-system in terms of which to grasp and evaluate this self-blinding fact. The Extremists, communist and fascist, could not conceive that class cooperation, race cooperation and overriding respect for evidence exist in reality rather than as mere propaganda. The Centrists—Republicans or Conservatives, and Democrats or Laborites—could not, for the same reason, conceive of the genuine contempt, treachery and cynicism with which they are viewed and treated, as a matter of course and of policy, by almost all Extremists, Left and Right. Only personal, often-repeated field-experience inside both political cultures—such as I encountered in ten countries in the mid-Thirties—can, in the absence of Unified Science, make logo-centrism and patho-centrism a visible, observable, and formulable phenomenon.[71,91] Unless you are an Einstein, you have to have lived in the free-falling elevator, rotating room, and on the solid ground—and to have moved back and forth several times from one to the other— to acquire and codify the coordinate systems of people who have spent their whole lives in only one of these windowless systems.⁹⁵

Such is the new coordinate system of political science, and its resolution of the long festering crisis in political thought. When implemented in political practice, it will give the democratic ships-of-state a political compass by which to set their course more safely.

"If this mixture of pragmatism and positivism does involve an ethical doctrine—" Heisenberg remarked to Pauli, "and you are certainly correct that it does and that we see it at work in America and England—by what compass does it set its course? You have claimed that in the final analysis our compass must be our relationship with the central order, but where can you find such a relationship in pragmatism?" p. 217.³⁹ My answer is, nowhere in pragmatism; but in Unified Science it is called the *coaction compass*⁴⁰ and is shown in Figures 11, 12 and 16.

In Chapter IV, separately studied genetic and psychological data will be related to their sociological, anthropological and political-scientific counterparts. Together, they will be assembled into a single discipline coherent with, and relevant to, the biotic and abiotic sciences.

9. CLOSING THE CIRCLE: SUMMARY.

The Unified Science Chart at the end of this book sums up this chapter's basic concepts and relates its basic figures to each other. It also relates them to a representative array of historic scientific concepts, listed in the left-hand column, which it completes, generalizes, assembles and/or maps into the single discipline, Unified Science. Experience shows that this fold-out chart, mounted on the wall, can serve the reader in coordinating his study of the rest of this volume, as perhaps also in other of his professional pursuits.⁹⁶

It was Leibniz who predicted (fifth entry from the top, counting Mendeleev) that one day it would be discovered that the many scientists who had believed that they were working in separate disciplines had, unknown to themselves, actually been working on a single discipline. He predicted that this will happen when the characteristics which their data have in common are abstracted and represented geometrically, as the *Universal Characteristic*. And he foresaw that into this Universal Characteristic would be mapped, besides the traditional sciences, the following: jurisprudence, medicine, and metaphysics.⁶⁴ Does this not close the dangerously broken Circle?

The chart before you opens the way to the fulfilment of Leibniz's prediction by assembling the constructs for closing the Circle in the Twentieth Century's Two Modes of Thought.⁷²

In *The Breaking of the Circle,* Marjorie Hope Nicolson describes what the Circle of our thought was like in the sixteenth century, before it broke. Does she not, in so doing, come close to describing how it has now been closed in the 20th? "Man *was* in little all the sphere," she says. "As he grew and flourished, so did his world; as he decayed and died, so too his world. God's 'pattern' was eternally repeated in macrocosm, geocosm, microcosm. Man's head was a 'copy' of God and the universe, not only in its shape, but in its being the seat of Reason. Man, the epitome of God and the world, was rational; so were the world and the universe, into which God had imparted some of His own rationality. Each of the 'three worlds' had its individuality, yet each was involved with the others, and all partook of God. Only since the seventeenth century has the poet felt the necessity of bringing together what the shears of scientific philosophy cut apart." pp. 106–7⁹⁷

The constructs listed in our chart's left-hand column are but a minute sample of the immense array of data and theories cut apart by the shears of 17th, 18th, and 19th century scientific philosophy; constructs which can, and will shortly be, mapped into the Periodic coordinate system. Is it not perfectly clear that we have been working on a single discipline?

The web-of-life, a tiny part of which is shown in Figure 14b, is keyed to the Biotic region of the Periodic coordinate system, shown at the center of the Chart. These will permit the conservationists' computers to talk turkey, as American slang puts it, with our fellow citizen's blind, deaf and dumb technology, and set a rational, mutually acceptable course of action.

Keyed to the Cultural region of our coordinate system, at the bottom of the Chart, we are now drawing several sets of psycho-political coaction webs: one of objective coactions, the others of subjective coactions. One of these is a web of the subjective political-economic coactions of Left-Center Liberals; one of Right-Center Conservatives; a third of Communist, and a fourth of fascist extremists.

The objective is to move our discussions from the barricade to the blackboard, to exchange our machine guns and bombers for chalk and erasers Or, in the milder terms of the Creative Center, the goal is to meet the need, as L. J. Livesey puts it, "Of institutions that seek to establish an innovative image while at the same time yearning for traditional respectability; and the never-ending process of deciding how much freedom one must yield as the price for community and government support . . ."[98]

Fulfilment of these objectives is inherent in our execution of Leibniz's project for, as he predicted would happen once his Universal Characteristic had been created: "If someone disagreed with me [on any subject, abiotic, biotic, or cultural] I should say to him, 'Sir, let us calculate!' And by taking to paper and ink we would settle the question."[64] Why is this possible? Because we have now come Full Circle: The natural law and moral law are recognized once more, and have been proved in the new 20th century terms to be identical.

NOTES AND REFERENCES

1. The original form of this concept was formulated jointly by W. V. Quine, H. G. Cassidy and E. F. Haskell in 1964, and has been developed by us to the present state. For a rich discussion and bibliography on related concepts of hierarchy, see Donna Wilson[2] (below).
2. Wilson, Donna, "Forms of Hierarchy; A selected Bibliography" *General Systems* Vol. **XIV,** 1969 (pp. 3–15) Yearbook of the Society for General Systems Research 2100 Pennsylvania Ave., N.W., Room 818 Washington D.C. 20006.
3. Mayr, Ernst, "Discussion Footnotes on the Philosophy of Biology".
4. Weaver, Warren, "Science and Complexity," in *The Scientists Speak* (W. Weaver, editor). Boni and Gaer, New York, 1947.
5. "I am the Alpha and the Omega, the beginning and the end." (The *Bible*, Revelation 21:6.) This sentence is herewith completed in a scientifically meaningful way which was not possible, when it was written, for Man to understand. It is also in accord with the other two references: "I am the Alpha and the Omega, the first and the last, the beginning and the end." (Revelation 22:13.) And "I am the Alpha and the Omega, saith the Lord God, who is and who was, and who is to come, the Almighty." (Revelation 1:8.) As the two limits of the Periodic coordinate system, and thus of Unified Science, this concept agrees with, and completes Teilhard de Chardin's scientific concept of Omega.[16]
6. A still more fundamental entity, the *parton*, has just been discovered. If verified, it will be classed in Major Period 0.
7. Haskell, Edward F., "Mathematical Systematization of 'Environment' 'Organism' and 'Habitat' " *Ecology*, Vol. **21,** No. 1 Jan. 1940.
8. Greenstein, J. and M. Schmidt "The Absorption Fine Redshifts in Parkes 0237-23," *Astro-physical Journal* (Correspondence) Vol **148,** April 1967.
9. The term *Major Period* is a conceptual extension of the term *Period* in Mendeleev's Periodic Table. Similarly, the term *Major Stratum* is an extension of the general term Stratum, coined to include the atoms' nuclear and electron shells and their analogues in the other sciences. The Major Periodic table appears below in a simple geometric form. The term, *Major* Period, was coined by my research assistant, Paul Mankiewicz. (I had been using a poor term, *Macro* Period.) A natural kingdom is a Major Stratum; a natural empire is a Major Period. (See Figure 2-4.)
10. The Biosphere, *The Scientific American*, September issue, 1970.
11. Periodic tables of molecules and geoid systems have not been constructed. But Stratification, Periodicity and Grouping are postulated for them, and the construction of corresponding Periodic tables is predicted.
12. Pirie, Norman W. "The Origins of Life," *Nature* **180,** 886–888, 1957.
13. Cassidy, Harold G., "The Kingdom of Biopoetic Systems-Phylogeny of the Cell." A chapter in *Unified Science—Assembly of the Sciences Into a Single Discipline,* Edward Haskell. Offset-printed in 50 copies by the National Institute of Health 1968, Xeroxed in 50 copies by the IBM Systems Research Institute, 1969.
14. Reinhold, Robert, "Scientists in Varied Fields Join in Attacking National Problems." *New York Times*, Oct. 6, 1970, page 1ff.

15. Anderson, O. Roger, *"An Interdiciplinary Theory of Behavior,"* Jour. of Research in Science Teaching Vol. **6,** No. 3, 1969, pp 265–272.
16. Chardin, P. Teilhard de, *The Phenomenon of Man*, (transl. B. Wall) Harper, New York 1959.
17. Murdock, G. P., *"Ethnographic Atlas"*, *Ethnology*, Jan. 1962–to date.
18. Hobhouse, L. C.; G. C. Wheeler, M. Ginsburg, *The Material Cultures and Social Institutions of the Simpler Peoples; an Essay in Correlation.* Chapman, London, 1915.
19. Toynbee, Arnold J., *A Study of History* (Somervell Abridgement of Vols. I–VI) Oxford Univ. Press, New York 1947.
20. Posin, Daniel Q., *Mendeleyev, The Story of a Great Scientist*, McGraw-Hill, New York, 1948.
21. Seaborg, Glenn T., *From Mendeleev to Mendelevium—and Beyond* Monograph presented at the Robert A. Welch Foundation Conference on Transuranium Elements—the Mendeleev Centennial, Houston, Texas, released by the U.S. Atomic Energy Commission, Washington D.C., Nov. 17, 1969.
22. From *Science Restated—Physics and Chemistry for the Non-Scientist* (page 300) by Harold G. Cassidy, Freeman-Cooper, San Francisco, 1970. Reprinted by permission of the publisher.—I have shifted the position of hydrogen (H) from Group IV to Group VII for reasons to be set forth, and have reversed the order of the Periods: We start at the bottom and count upward.
23. Latil, Pierre de, *Thinking by Machine—A study of Cybernetics* Houghton Mifflin, Boston, 1957. Originally, *La Pensée Artificielle* Gaillard, Paris, 1956; Sidgwick and Jackson, London, 1968.
24. Haskell, Edward F., "A Clarification of Social Science," *Main Currents in Modern Thought*, **7,** 45, 1949.
25. Adam, Charles Ernést, *Descartes; sa vie et son oeuvre*, Boivin, Paris 1937.
26. For a description of the role played by dreams in the development of mathematics and scientific theories, see Jacques Hadamard, *An Essay on the Psychology of Invention in the Mathematical Field*, Dover Publ., New York, 1945.
27. They appear in *Unified Science—Assembly of the Sciences Into a Single Discipline*; offset-printed by the National Institute of Health, 1968 and xeroxed by the IBM Systems Research Institute, 1969. (Three volumes, when completed.)
28. Haskell, Edward F., with preface and a chapter by Harold G. Cassidy, *"Unified Science—Assembly of the Sciences Into a Single Discipline"* Vol. I, *Scientia Generalis*. Offset-printed by the National Institute of Health 1968, xeroxed by the IBM Systems Research Institute, 1969.
29. This corresponds to the *Great Chain of Being* the medieval precursor of the System-hierarchy.
30. Gause, G. F., *The Struggle for Existence*, Williams and Wilkins, Baltimore, 1934, Hafner, New York, 1964.
31. Gause, G. F. and A. A. Witt, "Behavior of Mixed Populations and the Problem of Natural Selection", *American Naturalist* **69,** 725, 1935.
32. Wheeler, J. A., "Our Universe: The Known and The Unknown." *American Scholar*, Phi Beta Kappa Society, Spring, 1968.
33. Cassidy, Harold G., *Science Restated—Physics and Chemistry for the Non-Scientist*, Freeman, Cooper, San Francisco, 1970.
34. Wu, C. S., "Subtleties and Surprises—A Brief History of the Theory of Beta Decay." *Columbia University Forum* **9,** 1 (1966).

35. It has to be hyper-spatial because it corresponds structurally to mutually exclusive phenomena. The Periodic and Inverted-Periodic coordinate systems can no more be superimposed (occupy the same space) than can pro-particles and anti-particles. Their corresponding axes are oppositely directed. Whether their radius angles should be directed oppositely, as shown, or similarly directed I do not know. I have decided it arbitrarily, pending more knowledge.
36. Snow, C. P., *The Two Cultures—and the Scientific Revolution* Cambridge Univ. Press, New York, 1959.
37. Couturat, Louis: *La Logique de Leibnitz*, Olms, Verlagsbuchhandlung, Hildesheim.
38. Quine, Willard V., *Ontological Relativity and other essays*. Columbia Univ. Press, New York, 1969.
39. Heisenberg, Werner, *Physics and Beyond—Encounters and Conversations*. Harper and Row, New York, 1971.
40. Haskell, Edward F., *The Coaction Compass*—A general Conceptual Scheme Based Upon the Independent Systematizations of Coaction Among *Plants* by Gause, *Animals* by Haskell, and *Men* by Moreno, Lundberg, Horney and Others.—Mimeographed, 1948. (Mentioned in *Science*, Sep. 3, 1948 (p. 264).
41. Haskell, Edward F., "The Religious Force of Unified Science," *Scientific Monthly* 54, **545,** 1942.
42. In his review of *Physics and Beyond*, Elting Morrison remarked that "Events in the physical world take place the way a man thinks"[99]. And one of the great questions of philosophy has been how this could have come about. Unified science explains it by extending Bohr's principle of complementarity up from geoid systems (to which Newtonean physics applies) through the kingdoms of plant and animal ecosystems to that of human cultures. This explains why Man "thinks the way things happen in the physical world."
43. This term was suggested by W. V. Quine in discussions following this symposium: *Entropy*, he pointed out, is Greek for *turning in;* the opposite term should therefore be the Greek for *turning out*, namely *ectropy*. Since the coaction cardioid literally turns in and out of the circle of reference, his short, elegant term has been adopted and used throughout.
44. Weaver, Warren, "Science and Complexity," *American Scientist* 36, 537–44, 1948.
45. Eblen, William R., *Total Education in the Total Environment*, SPRED, Norwalk Conn., **18,** 1971.
46. *The Bible*, Paul's Epistle to the Galatians 6:7 "For whatsoever a man sows that he will also reap." RSV. 2.
47. *The Bible*, Matthew 7:12 "As you wish that men would do to you, do so to them; for this is the law and the prophets." Luke 6:31 "And as ye would that men should do to you, do ye also to them likewise."
48. Langlois, T. H., *A Study of Small-Mouth Bass, Micropterus Dolomieu (Lacepede) in Rearing Ponds in Ohio*. Ohio State Univ. Studies, Oct. 1936.
49. Haskell, Edward, "Assembly of the Sciences into a Single Discipline," *The Science Teacher* Vol. 37, No. 9, Dec. 1970, Supplement.
50. Wiener, Norbert., *Cybernetics—Or Control and Communication in the Animal and the Machine*, Wiley, New York 1948, Hermann, Paris 1948.
51. Bertalanffy, Ludwig von, *General Systems Theory—Foundations, Development, Applications*, Braziller, New York, 1968.

52. Bertalanffy, Ludwig von, "General Systems Theory—A Critical Review", a chapter in *Modern Systems Research for the Behavioral Sciences*, Walter Buckley, ed., Aldine, Chicago, 1968.
53. The paper in which these concepts were first expounded was called *The Coaction Compass* (see Note 40, above). Heisenberg's statement that "The problem of values . . . concerns the compass by which we must steer our ship if we are to set a true course through life" appears to infer this sort of thing. (See Note 39, above).
54. Coordinates of position begin at the origin and extend outward just as in the Cartesian system. But in the Periodic system, negative coordinates are subtracted from the zero circle. Example: Let the zero circle's radius be 15 units long. The coordinate X-5 is obtained by subtracting 5 units inward from the circle, and is thus a positive number: 10. This is the meaning of *relative* minus. The method of calculation was put forward by Gause[30] and Gause and Witt,[31] and was then generalized.[28] For another method of calculation, see the Addendum to Chapter 1.
55. Discussion of the lanthanide and actinide elements is here omitted, as they themselves are in Figure 5.
56. Location of Group 0 in quadrant 3 is neither geometrically necessary nor significant, since the 0 circle falls in all the quadrants equally. It is, however, empirically convenient, since Group VIII does fall in the $(-, -)$ quadrant, and Group 0 elements alternate with those of Group VIII in all the double Periods.

    Techniques have recently been invented by which inert elements can be induced to combine with others, something they never do in nature; that is, spontaneously. This shows that Period 6 Man need not resign himself to acceptance of apparently immutable natural conditions, which lower Period peoples could not even dream of questioning. We shall return to this crucial fact in later chapters.
57. Pauling, Linus, *College Chemistry, An Introductory Textbook of General Chemistry*. W. H. Freeman, San Francisco, 1951 (Chapter 31).
58. Simple peoples, and the half-conscious and sub-conscious parts of our own modern selves, tend to explain azoic phenomena by only partially or symbolically correct zoic and human analogies called *myths* and *symbols*. One-field specialists tend to react so strongly to avoid this error, that they also avoid correct analogies or veiled truths, and thus resist synthesis. This has resulted in the Two Cultures and widespread loss of self-determination through-out the West.—Unification of the sciences should overcome this paralysis and help us gain some control over our destiny. (See Jung, Carl *Man and His Symbols*, W. H. Allen, London, 1964).
59. Langer, Suzanne, *Philosophy in a New Key—A study in the symbolism of reason, rite and art*. Harvard Univ. Press, Cambridge, Mass, 1957.
60. Bohr, Niels H. D., *Atomic Physics and Human Knowledge*, Wiley, New York 1958.
61. Kuhn, Thomas S., *The Structure of Scientific Revolutions*, Univ. of Chicago Press, 1962, Chapter VIII.
62. There are, however, hundreds of isotopes which also map into the Periodic table.
63. The most famous approach to date has resulted in the models described in *The Limits to Growth*, produced under the auspices of the Club of Rome. Universe Books, N.Y., 1972.

Generalization of Mendeleev's Periodic Table 85

64. *Leibniz-Selections*, Ed. P. P. Wiener, Scribners, N.Y. 1951.
65. These "improvements and further developments" of the Periodic Law appear to consist in the substitution of *coaction* for *atomic weight* or *atomic number*, and in the substitution of the general geometric form in place of its originally concrete empirical form.
66. The decision as to which of these coactions obtains depends upon which organism is at the time the work component (x), and which the controller (y). These relations are complex, and are described and defined in more detail elsewhere.[28]
67. Subsuming by this and all other scientific classifications is W. V. Quine's deductively formulated theory of natural kinds.[38] His construct is therefore related to this work in the concluding chapter.
68. Bateson, Gregory, *Naven, A Survey of the Problems Suggested by a Composite Picture of the Culture of a New Guinea Tribe, Drawn from Three Points of View*. Cambridge Univ. Press, Cambridge, 1936.
69. Haskell, Edward F. *Switzerland's Vertical Front—The Migros Federation of Cooperatives in the Light of Systematic Social Science*. A chapter in *Gottlieb Duttweiler* by 65 authors. Speer Verlag. Zurich, 1948.
70. Baltzell, E. Digby, *The Protestant Establishment—Aristocracy and Caste in America*, Random House, N.Y. 1964.
71. Haskell, Edward F., "Geometric Coding of Political Philosophies." *Proceedings of the Second International Congress for the Philosophy of Science*, Vol. IV, Zurich 1954, Editions du Griffon, Neuchtâel, Switz., 1955.
72. Conant, James B., *Two Modes of Thought*, Trident, New York, 1964.
73. "Power grows out of the muzzle of a gun." Mao Tse Tung. "All of (recorded) history is the history of class conflict." Marx, Engels. (Analysed geometrically in Chapter V.)
74. Talmon, J. L., *The Origins of Totalitarian Democracy*, Praeger, New York 1952.
75. Djilas, Milovan, *The New Class—An Analysis of the Communist System*, Praeger, New York, 1957.
76. Oliver Cromwell, it has been pointed out to me, had his king beheaded, too. The crucial difference, however, is that Cromwell did this not gladly but reluctantly; that he steadfastly refused the crown; that he did not destroy the Minority; and that, when he bowed from the stage of history, he had markedly restored and improved his nation's over-all condition, which has maintained its form ever since.
77. Marx, Karl, and Friedrich Engels, *The Communist Manifesto*, London, 1848.
78. Until the sciences' theories are unified, *strength in either of these modes of thought requires and produces weakness in the other:* The parts of the universe are interlocked and display coherent structure. Until this is grasped, suitable concepts to describe it are formulated, and inter-disciplinary language-fouling is reduced, accurate empirical descriptions of its various parts necessarily result in an array of incoherent theories. Conversely, as long as the sciences remain discrete, coherent theory can be formulated only in non-empirical terms, and its content must contradict or ignore many empirical data and relationships. *To the extent that unified science is empirically correct and logically consistent, it reconciles these here-to-fore mutually exclusive modes of thought.* See Chapter V.

79. Strausz-Hupé, Robert; W. R. Kintner, J. E. Daugherty; A. J. Cottrall; *Protracted Conflict—A Challenging Study of Communist Strategy*. Harper & Bro·s New York, 1959.
80. Philberth, Bernhard, *Christliche Prophetie und Nuklearenergie*, R. Brockhaus Verlag, Wuppertal W. Germany 1964.
81. Koestler, Arthur, *The Sleep Walkers, A History of Man's Changing View of the Universe*. MacMillan, New York 1959.
82. It is no accident that the Extremist ideologists so strongly and consistently oppose all religions, and vice versa: Their dominant value-premises are diametrically opposite: the ideologists prize class or race conflict, the people of religion prize cooperation.
83. Benedict, Ruth, *Patterns of Culture*, Houghton Mifflin, Boston, 1934.
84. Sartre, Jean Paul, *The Philosophy of Jean-Paul Sartre*, (selections), R. D. Cumming, ed., Modern Library, New York, 1966.
85. Toynbee, Arnold J., *An Historian's Approach to Religion*, Oxford Univ. Press, London, New York, Toronto, 1956.
86. Robinson, John A. T., *Honest To God*, SCM Press, London 1963, Westminster Press, Philadelphia 1963.
87. This insight has been extended in *Divine Principle and its Applications by Young Oon Kim*, The Holy Spirit Association for the Unification of World Christianity, 1611 Upshur St. N.W., Washington, D.C. 20011, 1968.
88. Allardyce, Gilbert, *The Place of Fascism in European History*, Prentice-Hall, Englewood Cliffs, New Jersey, 1971.
89. Padover, Saul, K., editor, *Thomas Jefferson on Democracy*, New North American Library of World Literature, New York, 1946.
90. Geometrically, this coaction falls in quadrant 3, where the coaction cardioid displays two lobes and an apex. These geometric features correspond to the three sub-Groups in chemistry's Group VIII (Figures 5 and 11). The cardioid's negative apex $(- = -)$ is the geometric locus of the Sub-Group whose political expression is Nihilism. (Sometimes it is misnamed Anarchism.) This coaction is so destructive that its exponents frequently destroy themselves. See *Zero* by Robert Payne, John Day, New York, 1950. The "political spectrum" cannot map this phenomenon because it falls next to, yet between, its limits.
91. Haskell, Edward F. and Harold G. Cassidy, "*Plain Truth—and Redirection of the Cold War*," (136 pages) Privately printed and distributed, 1961.
92. Adorno, T. W. et al, "*The Authoritarian Personality*," Harper, New York 1950.
93. Communist and fascist regimes have the following six concrete traits in common: an ideology; a single party, typically led by one man; a terroristic police; a communications monopoly; a weapons monopoly; and a centrally directed economy. See Carl J. Friedrich and Zbigniew K. Brzezinski, "Fascism as Totalitarianism: Men and Technology." To these, Hannah Arendt adds a seventh common trait: "a foreign policy . . . directed toward world domination".[88]
94. Haskell, Edward F., "*Lance—A Novel About Multi-cultural Men*." John Day, 1941.
95. I have, however, expressed these two and several other centrisms—logo, patho, ego, ethno, and trato- centrisms—in fiction: "*Lance—A Novel about Multi-cultural Men*"[94]. Fiction permits the depiction of emotions which science strives to cancel out by means of invariants. Neither of the Two Cultures can

be supplanted by the other. A would-be C. P. Snow must be both scientist and novelist.

96. A less complete version of this chart about four meters high formed the visual background of the 1969 Boston symposium, whose papers (two of them in expanded form) are assembled in the present volume.

97. Nicolson, Marjorie Hope, "*The Breaking of the Circle*—Studies in the Effect of the 'New Science' upon Seventeenth Century Poetry." Northwestern Univ. Press, Evanston, I11., 1950.

98. Livesey, L. J., *Tomorrow's Education*, Program of the World Future Society's First General Assembly, Washington, D.C., May 1971.

99. Morrison, Elting G.; review of "Physics and beyond—Encounters and Conversations," *New York Times Book Review*, Sunday Jan. 11, 1971.

JERE W. CLARK: AN INTRODUCTION

Edward Haskell

Interchangeable machine parts that can be assembled were first envisaged and made in Connecticut. In 1800, Eli Whitney of Connecticut amazed and delighted Thomas Jefferson by staging a demonstration of parts assembly before a group of United States Government officials. Whitney casually dumped piles of meaningless parts out of several bags. He then picked them up at random from the piles, assembled complete muskets like rabbits conjured out of a magician's hat, and handed them to his astonished audience. And, to their amazement, every one of them worked!

This industrial feat made Connecticut so famous that when Mark Twain wanted to dramatize America's industrial advances over the Old World, the Yankee he chose for King Arthur's Court was a Connecticut Yankee. And rightly so, for among the most necessary conditions of industrial civilization is this abstract capability: the capability of machine parts—from the parts of transistors to those of planes, cars and houses—to be *assemblable.*

Now, there is no way around the fact that the author of Chapter III, Jere Clark, is not exactly a *Connecticut* Yankee. No matter what he's saying, whenever he opens his mouth he proclaims himself, loud and clear, to be from the South. (He once told me that the only Connecticut State College he could possibly have accepted an invitation to was *Southern* Connecticut.) So I'm introducing that interesting new type, the *Southern* Yankee. And the most interesting thing about this one is that while Eli Whitney, the first Connecticut Yankee, made interchangeable *machine*-parts, the second one, Jere Clark, has organized a Center for making *mind*-parts. For assemblable mind parts are just as vital to our new scientific and academic revolution as assemblable machine parts were and are to the industrial revolution.

You know what the modern university has become: a tremendous industry composed of a lot of high-powered departments making parts of mind; but parts which *have not been designed to be assembled*! (Figure IV-9.) The product of each department has his own special vocabulary and notation, and even his own world view, centered upon his particular ology. But each departmental specialist, characterized by his own logocentric background theory, is so remote from his colleagues in other logocentric fields that Dr. Clark talks about "interdisciplinary space", and likens unified science to a space ship for inter-disciplinary space travel!

And he does not just talk: he acts. Dr. Clark is perhaps the most active promoter of intellectual space-ship assembly plants known to history: He is Director of the Center for Interdisciplinary Creativity at Southern Connecticut State College. For the last three years he has served, and is still serving, as the international Chairman of the Education Committee of the Society for General Systems Research. He is past Coordinator of the Northeastern States Division of that Society, Executive Director of the Consortium on Systems Education in New Haven; U.S.A. Representative, Education Coordinating Committee, World Organization for General Systems and Cybernetics; and President of the Conneticut Chapter of the World Future Society. Dr. Clark is a charter member of the Leadership Council of the Buffalo-based Creative Education Foundation and is currently serving as a consulting editor of the *Journal of Creative Behavior*. He is also a consulting editor for *International Associations*, and Associate Editor of *General Systems Bulletin*. During the U.N. International Education Year (1970), he served as Systems Creativity Editor for *International Associations*.

In the beautiful, hilly outskirts of New Haven, at Southern Connecticut State College, Dr. Jere W. Clark directs his pilot plant for assembly of the sciences' now interchangeable parts. History thus repeats itself, as usual, on a higher level: what Whitney's plant for interchangeable *mechanical* parts was to our Industrial revolution, Clark's Center for Interdisciplinary Creativity is to our Scientific Revolution: an assembly plant in which to pull this civilization's unassembled *mind* together and gain, under God, control over its destiny.

Chapter III

The Role of Unified Science in Vitalizing Research and Education

JERE W. CLARK

1. CAPABILITIES OF UNIFIED SCIENCE MODELS

Designing a Vehicle for Mental Space Travel

It is common knowledge that the outstanding achievement of man to be recorded in history for the year 1969 is expected to be the consummation of interplanetary space travel. This year has indeed been a great year not only for the four moon walkers but perhaps even more so for the designers, producers, testers, launchers, and controllers of the space vehicles involved in these ventures.

Today—four days before the end of the year, 1969— we have met to test in a preliminary way, and to consider launching, a still more powerful and important kind of vehicle for space travel. This is a vehicle for what might be called "interdisciplinary mental space travel." Indeed, this vehicle is an interdisciplinary conceptual model whereby a specialist's mind can take flight to, and land on, distant intellectual planets, and return laden with a cargo far more valuable than moon rocks or gold dust. Although this special kind of space vehicle is not so romantic as those used by the moon walkers, it is far more important to the destiny of man—and much less costly.

Unique Capabilities of Mental Spacecraft

Fortunately, our mental space vehicle can have built into it a number of additional capabilities which the physical spacecraft do not have. As a stepping stone into the question of how an interdisciplinary conceptual model of unified science can vitalize education, we might note some of these unique qualities.

1. *Exploration of Social Space*

One of these extra capabilities of this conceptual model is its capacity to explore the galaxies of social space (as well as the galaxies of physical space). This model can link the various social disciplines with each other and with the natural science disciplines.

2. *Restructuring Knowledge*

The versatility and maneuverability of this vehicle is indicated by the fact that it can interlink sub-atomic particles, atoms, molecules, cell ecosystems, plant ecosystems, animal ecosystems and human cultures into coherent and orderly patterns. In so doing, the model will provide a basis for restructuring the entire realm of knowledge—thereby simplifying and vitalizing both education and research. As Russel L. Ackoff has put it, "We must stop acting as though nature were organized into disciplines in the same way that universities are."[1] Elsewhere in the same article, Professor Ackoff identifies the type of integration of knowledge which is required.

> In most problems involving organized man-machine systems each of the disciplines we have mentioned might make a significant improvement in the operations. But as systems analysts know, few of the problems that arise can adequately be handled within any one discipline. Such systems are not fundamentally mechanical, chemical, biological, psychological, social, economic, political, or ethical. These are merely different ways of looking at such systems. Complete understanding of such systems requires an integration of these perspectives. By integration I do not mean a synthesis of results obtained by independently conducted unidisciplinary studies, but rather results obtained from studies in the process of which disciplinary perspectives have been synthesized. The integration must come during not after, the performance of the research.[2]

Or, in the words of Sir Julian Huxley: " . . . we need a science of human possibilities, with professorships in the exploration of the future . . . the integration of science with all other branches of learning into a single comprehensive and open-ended system of knowledge, ideas and values relevant to man's destiny."[3]

3. *Providing a Meta-Language*

Another extra capability of this mental space model is that of providing the conceptual basis for developing a meta-language

The Role of Unified Science

which would be "spoken" in any galaxy in the universe of knowledge. Because of the technical nature and strategic importance of this semantic development, it will be discussed at length in a later section of this paper. It will suffice here to cite one testimonial to the need for it at the practical level. The relevance and importance of a functional meta-language are illustrated in personal correspondence from Mr. A. J. N. Judge, Assistant Secretary General, Union of International Associations, Brussels, to the author, November 6, 1969. Within the context of describing some of the special difficulties of establishing and operating a world management information system, Mr. Judge comments as follows:

> The area in which this work interacts with your own is that we hope to ensure that any success in your meta-language investigations, or similar developments in other countries, could be used as a means of relating terms indexing entities held within the system. Thus where two bodies may currently consider themselves as having no need to interact, we would hope to be able to show with the aid of work such as your own, in just which areas they could benefit from interaction.

4. *Crisis Prevention Management for Space-Age Democracy*

Another important capability of this mental transport medium is that it can provide a basis for crisis prevention as distinct from post-crisis administration of the process of educational change. Because of today's "explosion of explosions"—information, population, technology, interdependencies, pollution, crime, protests, and urban decay—this need is especially acute and seems to be destined to become still more critical in the immediate future. If we are to avoid letting trends toward complexity, rapidly rising expectations, and rapid change lock our society into an irreversible trend toward some highly regimented form of totalitarianism, we must somehow transcend the disciplinary barriers and create a new society. This new society would be based more on cooperation than on competition and, as is suggested in Figure 1, could make it possible to achieve ever higher and more meaningful forms of freedom.

5. *Problems Solved "Cheaper by the Dozen."*

Still another desired capability of the mental spacecraft is that of providing a scientific basis for the broad kind of scientific perspective which enables an entire scientific puzzle to be put together quickly.

94   *Full Circle*

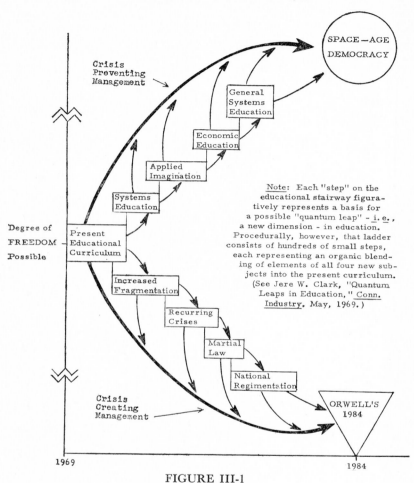

FIGURE III-1

This holistic vision is important even in putting together static jigsaw puzzles. It is still more important when all the pieces of a puzzle are in motion and are changing shape by the hour, as is the case in our nation's efforts to solve its major national problems such as poverty, urban decay, crime, unemployment, pollution, congestion and race prejudice. To continue to direct our efforts at each of these problems individually with a plan that is fairly independent of the plan for each of the other related problems is not only very costly and

The Role of Unified Science

ineffective, but is frustrating as well. Hence, it may be cheaper, quicker, easier, and more meaningful to develop a master, interdisciplinary plan for attacking these problems collectively, than it would be to solve any one of them by itself.

The futility of piecemeal approaches to our major problem is mirrored in the following excerpt from K. G. Harr, Jr., President of Aerospace Industries Association:

> We know much of what the future will bring in terms of problems. We know they will be big, complex, and serious . . . These problems represent the givens. We know they will be there—and we know they will overwhelm us if we do not find the means of coping with them. What we lack, thus far, is conviction that there is a means of getting hold of them. They seem so staggering in their size and complexity—so far beyond the capability of a single institutional segment of the community, public or private. . . And they are so interrelated that to proceed to try to solve any one of them in isolation from the other is often to create more problems than are solved by the effort. The dilemma thus presented has so far frustrated most efforts to come to grips with these problems. This condition of paralysis need not obtain. None of the . . . challenges lies beyond our already existing capacity for coping with them. The tools are already at hand; and included in those tools are not only the technological capabilities but experience in systems management and systems analysis as well as proven patterns of joint public and private effort.[4]

2. APPLYING UNIFIED SCIENCE MODELS TO EDUCATION

It is believed that Mr. Haskell's model for assembling the sciences is an example of a model which, in cooperation with supplementary concepts, models, and perspectives, has the potential of developing most—if not all—of the capabilities of vehicles for mental space travel enumerated in the preceding section. Much hard work, frustration, debate, reformulation, experimentation, testing, and adaptation will be required, however, before its potential can be realized.

Graduate Seminar

One of the highlights of the work of the Center for Interdisciplinary Creativity at Southern Connecticut State College was the sponsoring of Mr. Haskell's pioneering course, Assembly of the Sciences, in the spring semester, 1969. Twenty persons enrolled for either the three semester-hour credits or for auditor status. Included were graduate students, elementary, secondary and college teachers and administrators from the social and natural sciences, music, the humanities and fine arts. I am pleased to be able to say that I was one of the auditors.

The seminar is scheduled to be repeated at the SCSC Center in the coming spring semester. The following excerpts from the seminar syllabus will provide a general idea of the nature, methods, and purpose of the seminar.

> This seminar is designed first of all to provide an orientation to a newly developed, simplified approach to establishing functional communications bridgeheads between the social, biological and physical sciences, and the humanities and fine arts.
> To this end, the techniques of traditional specialization are extended to the task of assembling the basic data of the traditional sciences into a master or meta-scientific model. Analogous concepts and processes in different fields—e.g., a generalized form of the cybernetic process—are used as common denominators of all. The model serves as an intellectual road map to help the specialist in any field identify, reach, and interpret facts, principles, and processes which are especially relevant in other fields.
> The other principal purpose of the seminar is to apply this unified conception of science to the task of attacking systematically the major educational, social, political, and technical problems of our day.
> Such problems as deteriorating cities and ecosystems, crisis-ridden political systems, and obsolete educational curriculums, methods, and institutions will be related to student concerns.

Evaluation

The written and oral evaluations of the seminar were unusually encouraging. Generally the participants indicated that the seminar was believed to be several times as valuable as the average college course they had previously taken. Outputs which were revealed to be exceedingly valuable were:

    a. Improvement in attitudes toward science;
    b. Increase in degree of creativeness and open-mindedness;

The Role of Unified Science

c. Improvement in ability to transfer learning across disciplines;
d. Improvement in ability to understand literature in other fields;
e. Improvement in ability to cope with change;
f. Relevance to one's daily work.

It will be noted that the above factors are all critical in the present historical era. The fact that these are among the weakest outputs in most courses today further underscores the importance of the work in this seminar.

The course was thought to be almost equally valuable for school administrators, teachers, and graduate students. With appropriate modifications in the level of sophistication, it is believed that the seminar would be effective for college freshmen and some talented high school students.

Specific Applications

Several of the teachers are applying the basic ideas of the seminar in their own teaching and are reporting quite significant results in terms of student motivation and technical performance. The flexibility of the approach is suggested by the fact that teachers in many different fields and grade levels are using some adaptation of it. For instance, William Eblen and his associates in Wilton, Connecticut, use their own variation of this approach in their high school and college ecology project, Total Education for a Total Environment (TETE). Professor Rossalie Pinkham, Director of Laboratory Schools, Southern Connecticut State College, and Chairman, Consortium on Systems Education, New Haven, uses it as a springboard into, and as a frame of reference for, linguistic and social science subject areas. Chemistry teachers in high school use the periodic table as a springboard into interdisciplinary units. Biology teachers can use the general model as a functional framework for integrating the study of evolution in all the traditional sub-fields of biology and for relating evolution theory to psycho-social studies. Historians and anthropologists use it as a functional basis for explaining the process of change.

Being an economist who had already developed a broad economizing model for interpreting the universe of organized energy before meeting Mr. Haskell two years ago, I have blended his model into the economizing framework. A brief sketch of that master model will set the stage for describing the nature and importance of the task of developing a meta-language of the sciences, and for describing the particular approach we are developing at the SCSC Center for I-D Creativity.

3. ECO-CYBERNETIC COACTION THEORY

The synthesizing model of knowledge developed at the SCSC Center for I-D Creativity prior to the important link-up with Mr. Haskell's cybernetic coaction model is being called "eco-cybernetics." Eco-cybernetics fits the cybernetic communications apparatus into the broader framework of the generic means-ends or economizing process. By linking functionally the cybernetic flows of information with the economizing tasks of selecting aims, setting priorities, devising strategies, and identifying criteria for evaluation, it is possible to develop a more functional and simpler synthesis of knowledge in all disciplines. Thus we have a symbiotic, mutually-reinforcing relationship between cybernetic and economizing principles. Whereas cybernetics provides a basis for describing the patterns of inter-actions among the components of a system (and its environment), economizing principles provide a partial basis for explaining these patterns—why they emerge, what they are likely to lead to, and what alternative courses of action are available.

All of these economizing-cybernetic processes take place within the context of general ecology which includes human as well as natural ecology. Hence, the scope and purpose of general ecology are combined with generalized versions of the decision-making, economizing process of economics and the information control processes of cybernetics.

Several unique features of Mr. Haskell's coaction cybernetics make it superior to traditional cybernetics for our purposes. First, this coaction cybernetics is much broader in scope than is traditional cybernetics. It might be called cybernetics, "sub-cybernetics," and "supra-cybernetics." Cybernetics is the middle link in a chain of evolution from such sub-cybernetic (closed) systems as atoms and the "cybernetic-plus" systems such as human societies which have communications capabilities and operating characteristics which keep them from being considered to be cybernetic systems in traditional circles.

Second, by interpreting Mendeleev's Periodic Table in "cybernetic" terms, and then developing the cybernetic counterpart of that table in each of several other disciplines, Mr. Haskell has been

The Role of Unified Science

able not only to show the inter-relatedness of the various disciplines but also to express many of the key relationships in geometric terms. To think that the way may have been opened to express human values geometrically and in a way that can be related geometrically to other "values" is indeed remarkable.

A third unique and helpful feature of this coaction cybernetic model should be mentioned. That is the fact that it preserves the basic content and structure of each of the traditional sciences while adding the dimensions of general pattern and order, responsiveness and coordination. It mainly adds a specially "coded," conceptual lens for viewing the various sciences, and a more functional and operational way for approaching the various disciplines so as to interrelate them meaningfully and to convert more facts into usable knowledge.

This third unique feature may prove to be quite important in minimizing the amount and difficulty of changes a one-field specialist will need to encounter in up-dating his work. The specialist is not asked to forget what he knows and begin again, but rather to re-orient his image of his field.

4. UNIVERSAL META-LANGUAGE

We have noted in an earlier section of this paper that one of the real advantages of a general synthesizing model of knowledge is that it should make possible the development of a generalized language to use in the most basic aspects of many—ideally all—branches of knowledge. Such a language could be used to portray the functional operation of the most basic elements of all the social, biological, and physical sciences, and surprisingly large portions of the humanities and the fine arts as well.

The manner in which we at the SCSC Center for Interdisciplinary (I-D) Creativity are attempting to develop and apply the meta-language might be described figuratively in terms of two innovations from business and economic history.[5]

Developing an Intellectual Medium of Interdisciplinary Exchange

One of these innovations was the development of money in the economic world. The intellectual counterpart of money—our meta-language—is being designed to serve as a communications medium for the interdisciplinary exchange of ideas.

An emerging need for the conceptual counterpart of money can be seen by turning the pages of economic history back to the beginnings of organized efforts to develop and use a common monetary unit. As market places evolved historically to accommodate ever larger numbers of products, the need for some common unit of account or generally accepted medium of exchange was intensified. Likewise, as education and knowledge have developed to accommodate increasing numbers of scientific specialties, the need has increased correspondingly for an equivalent common means for communicating and exchanging ideas.

In other words, just as money can be used in industry to facilitate the inter-industry exchanges of goods and services, a meta-language, i.e. a "universal language" of science is being developed to expedite inter-disciplinary exchanges of ideas among research specialists.

This language will not only facilitate inter-disciplinary exchanges of ideas within the traditional framework of knowledge, but also should help significantly in creating a functional, operational synthesis of knowledge. This development is giving the specialist equipped with this language (along with supporting tools) mental wings to explore the whole universe of knowledge in search of particular lessons relevant to his own specialty.

Generalized forms of such economic principles as comparative advantage, diminishing returns and alternative cost are assumed to have just as much to do with the structure and functioning of an ant colony or a biological cell as they do with the operation of a business firm.

Likewise, such cybernetic concepts as input, output, sensor, controller, effector and feedback are assumed to have just as much to do with the operation of an economic market as they do with the functioning of the human nervous system or an engineering quality control process.

Guiding, Motivating, and Measuring Intellectual Effectiveness

The second innovation from the business world being built figuratively into our conceptual model was the development of profits which, in a theoretical competitive, free enterprise economic world, would serve to guide, motivate, and measure efforts to utilize scarce resources efficiently (effectively). Hence, our meta-language model is being designed to provide the intellectual counterpart of profits to guide, motivate and measure the results of efforts to be efficient or effective in allocating scarce education resources.

For decades, the educator has been aware of his need for more meaningful and objective criteria for allocating his intellectual resources. Traditionally, however, he has tended to treat this subject as Mark Twain said we treat the weather. That is, everyone talks but no one does anything about it.

This model is being devised to meet several requirements of a good testing, guiding and motivating vehicle of organization. The model must provide identifiable, measurable and demonstrable tests or yardsticks of intellectual efficiency broadly conceived. At the same time, it must provide a set of guidelines for increasing efficiency and a set of motivations leading individual persons and groups to strive to be more efficient. Furthermore, this efficiency generator, or model, along with the meta-language counterpart of money, must provide a basis for inter-relating all inputs and outputs of any given enterprise. It must help us identify the value of inputs by relating them to derived outputs. Finally, it must provide a basis for evaluating the results of experimental efforts and for utilizing these evaluations in designing follow-up experiments.

Although the present model relies heavily on intuitive judgements, it provides an operational basis for identifying the strategic variables and a means of organizing relevant information once it, or an estimate of it, is available. Research is being designed to focus a figurative magnifying glass and mental radar screen on each subjective value involved so as to measure it more carefully and/or to seek combinations of objective sub-variables which will yield more practical results. Value engineering principles are being combined with cost-effectiveness and cost-benefit analyses in an effort to determine more clearly what types of educational inputs are producing the desired educational outputs. As experimental progress continues to be made along these lines, the expectation is that more reliance can be placed on objective, measurable, demonstrable factors.[6]

Translating Ethnic Bi-Lingualism Into Scientific Pan-Lingualism

This discussion of meta-language would be incomplete without some extensive excerpts from a letter I recently received from Mr. Haskell. In the letter he was calling to my attention an unusually perceptive article, "Bilingualism and Information Processing," by Paul A. Kolers, in the March, 1968 issue of *Scientific American* (pp. 78—93) and was stating his interpretation of it. I am pleased to be able to quote the following excerpts from that letter.

> Each discipline has, and must have, a certain vocabulary of its own: thousands of words that refer to its data on the lowest level of abstraction. These represent the "concrete manipulable objects" mentioned on page 82 of Mr. Koler's article. (For psychology and the humanities they include emotions, feelings, values.) The point however is, first, that each discipline now has, but in most cases does not need to have, a set of words which represents the grouping, the classification of its low-level words. These are the "abstract words" in the reference above, comprising part of each discipline's vocabulary. These many diverse abstract vocabularies cause the disorganized complexity of modern education and thought, producing much of the confusion in the students' minds. They require a vast amount of memorization, conceal meaning, and prevent understanding.
>
> Our model of unified science corrects this situation: It accepts the data of each of the sciences and humanities, and the vocabularies which denote them. It also accepts many "game tree" classifications such as the taxonomic series, and all others which are logically compatible with each other. It then classifies these diverse sets of data and classifications in a single manner and vocabulary: The manner and vocabulary in which the chemical elements were classified by Mendeleyev and Maier a century ago, and which have since been improved by many others . . . (See fold-out chart and Figure IV-11).
>
> In unified science, the same set of abstract concepts occurs over and over, once for each discipline. This has the following effect: It makes learning vastly easier; it reveals the meaning of each higher set of data (as shown in our Periodic co-ordinate system) in terms of the lower sets; and the meaning of each lower set of data in terms of the higher sets. Thereby it increases the breadth and depth of the students' understanding.
>
> This is why our seminar students who answered your questionnaire found "Assembly of the Sciences" on the average 3.76 times as meaningful, relevant, and useful as the average graduate and undergraduate course they had taken.
>
> This American execution of the Royal Society's second objective is possibly more important than reaching the moon. It

The Role of Unified Science

facilitates and improves education here, and potentially around the world, three or four times over right at the start. It makes possible the inter-disciplinary research essential to the successful management of our complex system of ecosystems. It permits, as shown in your conference, the restatement of ancient religious

Proposal for a

JOINT COUNCIL ON UNIVERSAL SYSTEMS EDUCATION

MISSION

TO PROVIDE SIMPLY, INEXPENSIVELY, IMAGINATIVELY, AND OPERATIONALLY THE CONCEPTUAL CAPABILITY OF CONVERTING OUR FRAGMENTED COMPLEX OF DOMESTIC, DEFENSE, AND SPACE PROBLEMS INTO AN INTEGRATED PATTERN OF GLOBAL OPPORTUNITIES.

Strategy

To re-structure our educational system in terms of eco-cybernetic processes so as to enable it to provide an operational basis for developing the global perspective, the administrative ingenuity, responsiveness, and coordination required for peaceful living in increasingly dynamic, complex, and interdependent democracies.

Procedure

To develop creative systems-education programs which provide the mental sensitivity, flexibility, versatility, fluency, and dexterity required to transform the entire educational system into dynamic, functionally integrated, individualized patterns of self-programmed, adaptive behavior.

Programmed Sequence
of
Procedural Objectives

(Please see next page.)

FIGURE III-2

truths in modern cybernetic terms. And it successfully transfers political controversies from the barricade to the blackboard, as also demonstrated in practice elsewhere.

Unified science not only vitalizes education and research, but redirects dangerous religious and ideological cold wars into warm teamwork.

5. THE CHALLENGE BEFORE US

Against the background already sketched we can now consider the challenge before us. A part of the challenge is reflected in the following excerpt from Dr. N. Henry Moss' address, "The Pursuit of Knowledge—Synthesis or Fragmentation?", given as retiring President of the New York Academy of Sciences:

> Where are our great synthesizers of knowledge? . . . Industry, government, universities, and medical schools have need of and eagerly seek capable people who can effectively give the broad sweep, recognize the important from the unimportant, and maintain a reasonable understanding of the literature in multiple medical and scientific areas. . . . We must be able to develop a corps of such scientists and physicians as synthesizers and integrators to help guide us in directing our resources toward optimum use. . . . We must create this kind of fine generalist in the image of a captain of a team, a strong and vigorous leader with intelligent insight and technical know-how in multiple fields.[7]

Another part of the challenge is reflected in the sketch of Figure 3 which is a proposal for designing and developing a global network of centers and institutes for the promotion of general systems methods, philosophy, attitudes, and viewpoints. The terminology used there is suggestive rather than definitive. That mission profile is the embryo of a possible blueprint for a global campaign to translate the capabilities of unified science models into a widespread program of action. The assumption is made that the entire realm or universe of knowledge would be restructured as a prerequisite to any adequate reforms in curriculum and administration.

Another part of the challenge inheres in the global political setting in which both scientific research and education take place. Our national aspiration to send men to Mars and return them safely could be considered within this context. Because of the great economic and engineering capabilities required, we, as a nation, might do

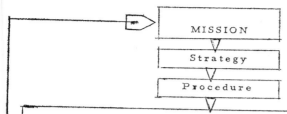

Procedural OBJECTIVES

1. DEVELOP BASIC ECO-CYBERNETIC MODEL AND META-LANGUAGE
   a. Develop an operational, organic, eco-cybernetic paradigm for functionally unifying progressively more facets of the sciences and humanities.
   b. Develop a general meta-language based on this paradigm.

2. DEVELOP APPLIED ECO-CYBERNETIC MODELS
   a. Intellect
   b. Behavior
   c. Learning Behavior
   d. Creative Learning Behavior

3. BLUEPRINT PROTOTYPE OF AN ECO-CYBERNETIC EDUCATION SYSTEM

4. ENCOURAGE DEVELOPMENT OF NATIONAL NETWORK OF U.S.E. CENTERS AND EDUCATIONAL SYSTEMS
   a. Implant prototype in a pilot community.
   b. Establish a Joint Council on U.S.E.
   c. Duplicate prototype center and education system throughout the U.S.A.

5. INTERNATIONALIZE THE MOVEMENT
   a. Main export of the U.S. economy
   b. Vehicle for international exchanges of products, ideas, and ideals.

Proposed by Jere W. Clark, Dir., Center for Interdisciplinary Creativity, Southern Conn. State College, New Haven, March 1, 1969. (Revised December 23, 1969.)

FIGURE III-3

well to consider teaming up with the Soviet Union for this venture. We know that the Russians are quite advanced in the development and application of broad-gaged cybernetic and other broad systems models to political, scientific, and economic ventures.

In view of these considerations, along with the extremely important capabilities of unified science models, might there be something this group gathered here might be able to do to dramatize the potentials for global progress, which might result from concerted cooperative action to launch some form of this mental space vehicle? Could we, for example, initiate action leading toward a dramatic proposal that the governments of the USA and the USSR jointly appoint a global task force for exploring the possibilities of joint exploration of both physical and intellectual space?

In other words, might we be able to use the dawning Seventies to initiate action which might be of even greater historic significance than the lunar landings earlier in the century?

NOTES

1. See "Systems, Organizations, and Interdisciplinary Research," *General Systems Yearbook*, Vol. 5 (1960), Society for General Systems Research (quoted in Anthony J. N. Judge, *The Improvement of Communication Within the World-System*, Union of International Associations, mimeographed, September, 1969, Appendix Figure 2, page 5).
2. *Loc. Cit.*
3. "The Crisis in Man's Destiny" in Donald E. Hartsock, ed., *Contemporary Religious Issues*, pp. 178-179.
4. *Harvard Business Review*, March-April 1967, page 10 (quoted in Anthony Judge, "Organizational Apartheid—Who Needs Whom in the Second United Nations Development Decade (1970–1980)?," Union of International Associations, Brussels, 1 rue aux Laines, Brussels 1, Belgium, UIA Study Papers INF/1, page 16).
5. This section is adapted largely from J. W. Clark, "Quantum Leaps in Education," CONNECTICUT INDUSTRY (Vol. 47, No. 5), May, 1969.
6. The general nature and content of this eco-cybernetic model are described in J. W. and J. S. Clark, eds., *Systems Education Patterns on the Drawing Boards for the Future*, Kazanjian Economics Foundation, 1969, chapters 6 and 7. Another article which provides additional background information is J. W. Clark, "Facing the Crisis of Intellectual Poverty," *Speech Journal*, Spring, 1968 (scheduled to be reprinted in a forthcoming issue of *The Journal of Creative Behavior*).
7. *Transactions of the New York Academy of Sciences*, January, 1968.

ARTHUR R. JENSEN: AN INTRODUCTION

Edward Haskell

The tendency of Right social political theoreticians to exaggerate the role of genetic factors in human cultures, and to minimize the role of habitat factors, and the converse tendency of Left political theorists to minimize the role of genetic, and to over-stress the role of habitat factors, were adumbrated in Chapter II.

In each case, these tendencies are shared by Moderates and Extremists. These groups agree in their biases, but differ enormously in the degree of violence and of falsehoods with which they denounce their opponents' biases. And among their opponents they count the scientists who have the courage to study this strategically important subject with regard for the accuracy and completeness of their findings, and disregard for the grinding of social axes within races or the goring of racial oxen within humankind.

Outstanding among these responsible modern scientists is Arthur R. Jensen. His studies of human intelligence are well formulated, empirical, careful, broadly verified. To these findings, Jensen relates the relevant spectrum of animal-intelligence researches independently conducted by M. E. Bitterman and his associates. Together, these studies comprise the empirical psycho-physio-ecological section of this two-part chapter.

Its first section consists of the theoretical Periodic Table of human cultures, based upon independent anthropo-sociological studies made by the many mutually independent workers cited and quoted in Chapter II.

The fact that these two independently developed sets of findings confirm and supplement each other testifies to the vigor of our civilization's Creative Center, defined and mapped in the Periodic coordinate system of Political Science, Figure II-16. We expect, of course, to be questioned and critized, to have mistakes or inaccuracies pointed out, and improvements suggested by our loyal opposition not only among scientists, but also among laymen of Center, Left Center and Right Center; by the positively oriented groups and individuals there defined. All these we welcome. We also expect to be attacked, derided, vilified, and character-assassinated by organizations and individuals of the Far and Extreme Left and the Far and Extreme Right as there defined, both inside and outside the multi-versities. We therefore call upon all honest men and women of the Center to stand together with us in calmly repelling and coolly refuting these destructive and disorganizing forces, and in redirecting them into constructive work.

The tables have now been turned in the psycho-social sciences. Up to this point they favored breakdown and disintegration; now they favor the genesis of a higher civilization. It is our duty to seize the initiative and press ahead to the offensive.

History has been turning these tables consecutively in the abiotic and biotic sciences, and has now done it in the cultural area of science. Consider three historic figures: What Copernicus did in abiotic science in the 16th century and Pasteur did in biotic science in the 19th, Arthur Jensen is probably doing in cultural science in the 20th century. Each of these scientists forced mankind to face new facts, and in their light to correct strongly held, crippling misconceptions. Copernicus helped to correct the popular astronomic and geographic world view that crippled navigation; Pasteur helped overthrow the deeply held biological ideas that had permitted millions to get sick; now Jensen is forcing us to examine bio-cultural views threatening our culture through massive deformations of its education.

The first two scientists brought, and Jensen promises to bring humankind enormous benefits. Copernicus and his peers vastly improved our navigation on land, sea, air, and now in space; Pasteur and his fellow workers cured millions and prevented more millions of people from getting sick; now Jensen and his associates are helping even more millions of people to avoid educational frustration and career disappointments by guiding them into rewarding and useful occupations.

We fully expect Jensen and his co-workers' contribution to be misunderstood and derogated, as were the pioneering works of those who went before him in other fields. But we intend to see to it that he gets a hearing. For only by listening to honestly presented empirical evidence, and testing it, has mankind found its way out of the difficulties it has, by its nature, always created for itself.

In September of 1969, concerted attacks were mounted against Jensen's careful, honest and important work under the pejoratively intended slogan of "Jensenism". I therefore invited Dr. Jensen, by long-distance telephone, to summarize his studies of human Stratification at the approaching A.A.A.S. symposium in Boston. He accepted, and the paper was announced thus in the program. As our correspondence developed, however, Dr Jensen realized that my object is to mobilize the masses of accumulated genetico-psycho-anthropo-social data into a Periodic Table of Human Cultures, and with this paradigm to challenge the self-perpetuating paradigm of egalitarian promiscuity and anarchic data collecting in the social sciences, as Mendeleev had challenged it in chemistry a century earlier. Jensen was in accord. But he was reluctant to assume responsibility for any work except his own. We therefore agreed to write this paper jointly: Its first part, which is primarily theoretical, by me; the chiefly empirical second part by Arthur Jensen. The result is the following chapter, of which my part has been considerably expanded since the symposium, with Dr. Jensen's knowledge and approval.

Part I by Haskell: *Anthropo-Socio-Historico-Linguistic Bases of the Periodic Table*. Part II by Jensen: *Direct Psychological and Genetic Empirical Basis of the Periodic Table*. Our joint paper, Chapter IV, is *Framework of the Periodic Table of Human Cultures*.

NOTES

1. Program, American Association for the Advancement of Science, Boston, Massachusetts, December 1969 (page 249).

Chapter IV

Framework of the Periodic Table of Human Cultures

EDWARD HASKELL

PART I: ANTHROPO-SOCIO-HISTORICO-LINGUISTIC BASES OF THE PERIODIC TABLE.

1. CULTURAL PERIODS, STRATA AND SUB-STRATA

The accumulated data of the bio-psycho-social sciences have the same two major aspects as those of the physical sciences. Namely, the quantitative, scalar aspect and the qualitative, directional aspect. In the second chapter's last part, we discussed the qualitative aspect, and mapped it into the framework of the Periodic coordinate system in Figure II-15. In this chapter we proceed to organize the quantitative, scalar aspect of the accumulated data—social, mental, and genetic—and to do so in terms of levels of organization as we did for abiotic and biotic systems in Chapters I and II.

Our paradigmatic example, Mendeleev's Periodic Table, seems superficially to be constructed in terms of increasing atomic weights or atomic numbers. What this means on the higher level of background theory, however, is increasing control-capability in the nucleus with corresponding increases in the organized complexity of the electron cloud, and of the atom, the system they jointly constitute. So also with the Periodic Table of Human Cultures.

Superficially, this table seems to consist of a growing hierarchy of tools, a growing food hierarchy, an increasing hierarchy of social Strata, of age-grades or Sub-strata within each Stratum, and a growing hierarchy of vocabularies.[1] What this means fundamentally,

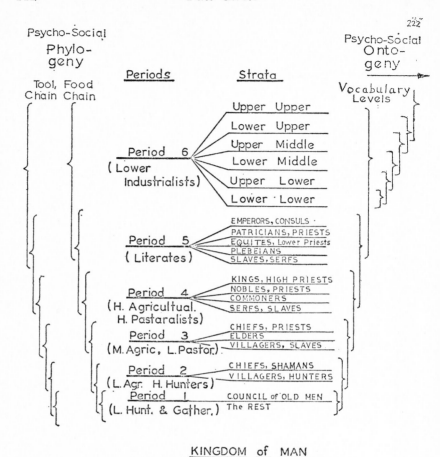

FIGURE IV-1   The kingdom of human cultures: Periods and Strata.

however, is increasing control-capability in each social system's controller, its leading Minority, with corresponding increases of production capability in its work component. These increases are manifested in the number of social Strata which comprise each Period's Majority and Minority, and growth of the dimensions (extensive and intensive) of the vast ecosystem they jointly organize.

In 1869 Mendeleev did not know that protons exist nor, of course, that his classification is actually based on the chemical elements' proton numbers.[2] He based his Periodic table on those existing data which most closely approximate the atoms' proton numbers; namely, their atomic weights. We, similarly, do not know the psycho-genetic structures which underlie our human Periodic Table.

What we know today is, that there are quantum-like discontinuities in our four hierarchies of data—tool, food, social, and linguistic—which occur together; that by postulating a hierarchy of psycho-genetic structures which accord with the principles of genetics, physiology, psychology and sociology, we can account for these data; and that by constructing our Periodic table in terms of this hypothesis, we obtain the same kind of organization of accumulated psycho-social data that Mendeleev obtained with the accumulated chemo-physical data of his time.

I predict that the postulated hierarchy of psycho-genetic human structures will be discovered empirically after our Periodic table's announcement in 1969, as were the atoms' nuclei and electron clouds, and their component particles, after the announcement of Mendeleev's table in 1869.

The two chains of interlocked braces on the left-hand side of Figure IV-1 represent the tool hierarchy and food hierarchy. The first Period's brace includes tools ranging from stone hand-axes to wooden boomerangs and spear-throwing sticks. And the foods are wild plants and wild animals, including insects. A few dogs help in hunting and in protection, and often also serve as food.[3,4] The society is a small band or horde, and consists of just one social Stratum, represented in the middle column. The work-component or Majority of the band is made up of the younger men and women, and the children; the controller or Minority is the council of old people; that is, people usually in their thirties. (Controller and work component are thus composed of Sub-Strata.) And there is just one vocabulary, in the sense which I will presently define.

This explanatory hypothesis is outlined in the psycho-genetic diagram, Figure IV-2. The first human Period is represented by a single characteristic number, as is the first Period in each of the other Periodic tables. The kingdom of man—which emerged from the natural empire of animal ecosystems, Major Period 6—is Major Stratum 7. Accordingly, its first Period has 1 Stratum and 1 Sub-stratum.

Turning back now to Figure IV-1, the second Period has two inter-locked braces: the lower one includes most first Period tools and foods, though usually somewhat modified; and the upper brace adds new ones: some seeds are planted and grown instead of just eaten; some small animals—such as pigs, sheep, fowls—are tended

114   Full Circle

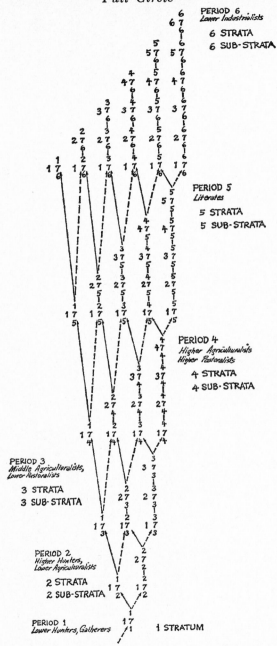

FIGURE IV-2   Phylogeny and ontogeny of human cultures.

† Note well that the Sub-strata or ontogenetic stages shown here are *human* abstraction levels. They *include*—and, in fact, *presuppose*—the mental levels of the highest animal Periods. Human Period 1, for instance, is shown to have a

Periodic Table of Human Cultures 115

and bred. New kinds of equipment for this agriculture are represented by the upper brace. The society consists of a few nomadic villages, and two social Strata.

Genetic Basis of Human Straticatifion.

Turning again to Figure IV-2, our explanation of how this came about is this: The first Stratum consists of the unmutated descendants of Period 1 people—though somewhat modified by the new foods and technology. (Unmutated descent is symbolized by the solid leftward arrows.) The much smaller upper Stratum, however, consists of a few descendants who have undergone a complex of so-called pleiotropic mutations, producing the psycho-physiological capability of conceptualizing and executing the new and more complex relations which we call *Lower Agriculture* and *Higher Hunting*. It also produces the capability of helping the unmutated relatives and neighbors, the Majority, to participate in, rather than to sabotage, these slow, difficult and often tedious operations. This new psycho-genetic set of capabilities is summed up and called (for now) *a higher level of abstraction*, though other factors are involved besides abstraction.[1]

This second human abstraction level appears phenotypically by way of an additional ontogenetic stage, and is symptomatized by what Johnson O'Connor calls a *higher vocabulary ceiling*, presently to be described.[5] (Mutated descent is symbolized in our figure by dotted rightward arrows.) Period 2 thus consists of two social Strata: the big, ancient, unmutated yet modified Majority, undergoing one human ontogenetic development-stage; and the small mutated Minority, undergoing two development stages. These are the two Sub-strata of Sub-stratum 2, and are labeled 1 and 2 respectively.

The postulated psycho-genetic process characterizing human

---

single Sub-stratum; but that is *in addition* to all of the highest animals' Sub-strata. This basic and strategic truth is recognized by all the great religions, but was deplorably ignored by Charles Darwin and most biologists since Darwin. It is officially ignored by Dialectical Materialists, who base their claim to being "scientific" on this and similar aberrations of the one-field specialists. In practice, however, human abstraction ceilings are carefully observed and utilized by Dialectical Materialists, as shown in their strategically graded levels of communication: *agitation, propaganda, official theory* (for intellectuals), and *secret theory* (for high Party leaders). It is by means of *agitation*, on the lowest level of human abstraction, that they incite what they call *the masses* against courageous, conscientious testers of intelligence. (See Chapters II and V.)

evolution is now clear: both of the second Period's Strata produce unmutated descendants, represented by the two solid leftward arrows. *Both of them also produce mutated descendants:* the first Stratum goes on mutating some descendants with the capability of two human abstraction levels. These join the unmutated descendants of Stratum 2 by what is incorrectly called just *social* mobility, though it is *genetico-social* mobility: *the capability* is inborn or genetic; the *opportunity of realizing this capability* is social, political, and legal.

Meanwhile, the second Stratum is producing its own mutations, Stratum 3 shown by the dotted arrow. This literally new kind of people displays a third human abstraction capability, generated by way of an additional ontogenetic development stage or Sub-stratum.

This new Minority invents the techniques and tools and institutions of Period 3, the Middle Agriculturalists and Lower Pastoralists. It adds these to the traditional tools and techniques, and to the Majority who use them, modifying some and eliminating others in the process. (This mutated innovator is the kind of son who, according to the so-called Oedipus myths, has to kill his father or has to be killed by him.[6] This psycho-genetic hypothesis makes such myths understandable, on other than sexual grounds.)

This is an extremely simple and straight-forward theory, as Figure IV-2 shows: Each human level of abstraction is characterized by the capacity to conceptualize, reflect upon and organize all preceding levels of thought and action. Abstraction levels are thus quantized, discontinuous jumps of capability. They are postulated to emerge by genetic mutation and to be transmitted genetically. Being genotypes, and thus potentials or capabilities, their phenotypic actualization occurs only in habitats whose highest vocabulary level and behavior level is equal to, or higher than, the genotype's (potential) abstraction ceiling.

Social Strata.

Each person's habitat is centered upon and controlled by his physical family or its cultural equivalent. This controller is surrounded by its friends and neighbors, who tend to display the same abstraction ceiling in their behavior and vocabulary. Collectively, this sociocultural habitat is called a social *Stratum* or social *class*. This habitat displays the same abstraction ceiling as that of its unmutated descendant, the above person in question. Genetic mutations, however, are constantly occurring, as are variations in habitat properties.

The Basis of Social Mobility.

Since some increase and others lower the offspring's abstraction ceilings, these variations produce the usual more or less normal distribution of abilities, with various ranges of spread. The result is an overlapping of Stratum actualities such that the most able members of a given Stratum equal or exceed the average of the next higher Stratum, and that its least able members fall below the next lower Stratum's average. This overlapping results in genetic-social mobility, upward and downward.

It follows that when mobility is blocked long and effectively, it results in anomalous relations between controller and work component, and thus in breakdown or disintegration of the system. It also follows that when mobility is artificially generated, forcing large numbers of unmutated, actually low-potential minds into control positions, the society transmutes down to the corresponding Period.

Stated more simply, each person is a complex key and every social habitat a compound lock. For thousands of years mankind has matched its human locks and keys by the expensive, painful, inconclusive method of theory-less trial and error. The first part of this chapter departs from this method by putting in train the classification of locks and keys. Its second part, by Arthur Jensen, describes the ever more accurate and reliable diagnoses of peoples' inborn genotypic capabilities. To this must then be added the diagnosis of habitat capabilities for transforming the individuals' genetic potentials into phenotypic actualities. Together, these operations will become a technology second to none in importance. (In the section on mapping the web-of-mind, a method is developed for cheap and painless computer simulation of the mutual consequences of placing each of the various kinds of students in any of the various kinds of schools, extant and theoretical. No means and effort should be spared to develop this technology as fast as possible.)

Human Stratification and Periodicity, Figure IV-1, and their development, Figure IV-2, are here, I believe, accounted for in a manner consistent with the data and operations of all the sciences involved: with genetics, psychology, linguistics, history, anthropology, and sociology.[9] Geometrized political science, briefly presented in the second Chapter which is strongly concerned with the qualitative, directional component of human cultures—accords with all these data and theories.[1] Its detailed presentation, however, like that of the present quantitative (not numerically, but geometrically quantitative) studies, display the same background structure as do the six lower Major Strata (natural kingdoms) and Major Periods (natural empires), conforming to what Heisenberg calls the *central order*.

The characteristic numbers in Figure IV-2 represent the cultural equivalents of biotic characteristic numbers, Figures II-14 and II-15: In the center position is humankind's kingdom or Major Stratum 7. Above it is the individual's or group's social class or Stratum number; that of its potential abstraction ceiling. At the bottom is their society's Period number, and at the left, the number of the individual's or group's actual, phenotypic abstraction level at the time in question; the number of its Sub-stratum or ontogenetic level.

Implicit in this abstraction hierarchy is the following exceedingly important fact: human beings are by definition unable to grasp abstraction levels higher than their own abstraction ceilings; unable to understand, often even to become aware of, and never to operate on levels of abstraction higher than their own abstraction ceilings. (Hence the loneliness, frustration, impotence, and sometimes martyrdom of great and good men and women throughout human history.) These characteristic numbers and some of their consequences appear below in the Periodic Table of Human Cultures and in our mappings of the web-of-mind.[8]

By postulating three more cumulative "repetitions" of this process, we obtain human Periods 4, 5, and 6 with the correct number of Strata and Sub-strata in each Period; the correct kinds of tools, foods, social positions, and vocabulary levels in each Stratum; the correct kinds and amounts of control; and the correct kinds and amounts of socio-genetic mobility among the Strata Skipping Periods 4 and 5 for lack of space, we come to our own Period 6, Lower Industrialists.

According to the most detailed study made thus far, the five-volume *Yankee City Series* by W. Lloyd Warner and his associates,[9] each of our Lower Industrial civilization's six Strata displays, among others, the following biological traits:

All these Strata are inter-fertile. Mankind thus constitutes a single species.

Human Strata, however, are highly endogamous. In Yankee City, during the period under study, 91% of all immediate families contained *no* person of any Stratum other than their own. (Just nine percent contained one or more persons of some other Stratum.) Such a degree of endogamy is characteristic of true sub-species. Each Period, after the first, thus constitutes an *ecocline*. This appears graphically in Figure IV-2 as increasing Period span.

The following data confirm this conclusion: longevity is directly correlated with Stratum number, and thus with the number of each Stratum's ontogenetic stages.[9] The higher the Stratum, the more development stages it has to undergo, and the longer it has to live to do so. And it *does* so.

Reproduction, and thus population-size, is *inversely* correlated with Stratum number, as is the age of marriage, and thus with the greater speed of maturation.[9] The older and lower the Stratum, the more it relies upon the older biological process of physical reproduction, and the faster it maturates physically.

Johnson O'Connor and his associates, moreover, have discovered and thoroughly verified the existence of five distinct vocabulary levels in American communities. "The rate of vocabulary acquisition describes a hyperbolic curve [Figure IV-3] and each level of functioning is a separate curve, limited by its own horizontal asymptote. Statistically, these curves are not broken. The individual is 'locked-in' to the pattern begun when he was a child."[5] There is, however, about as large a number of exceptions as mutation genetics would lead one to expect.

I predict that investigation will disclose a high correlation between these five vocabulary levels and the five lower socio-genetic Strata described by Warner and associates. This correlation is represented in Figure IV-1's upper right-hand corner by a series of interlocking braces. These indicate that the community displays a linguistic System-hierarchy; a *linguacline*. (This holds not just for Period 6 but for all human Periods except the first; but in ever lesser degrees.) This concept should greatly facilitate the very important study of linguistics, and be improved by it in turn.

As the braces in Figure IV-1 imply, I further predict discovery of a vocabulary level in the position and shape indicated by the dotted line I have drawn into Figure IV-3. This level will be displayed by the highest socio-genetic Stratum in Period 6. I also predict that *at least* one still higher level will be discovered. And finally that the new words which characterize each higher vocabulary level will be found to lie one level of abstraction higher than the highest of the preceding vocabulary ceiling. It follows that the communities of every human Period after the first will be

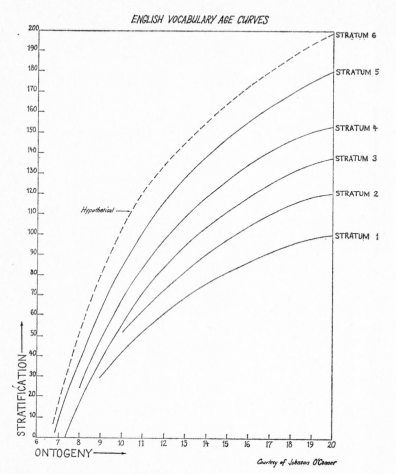

FIGURE IV-3 Stratification of English Vocabularies in the United States. By courtesy of Johnson O'Connor.[5]

found to constitute what I propose to call a *linguacline*; and that the human Periods' top Strata form one too.

Space limitation permits mention of just one more item in the enormous array of evidence supporting our assembly of these data. Each Stratum's number is directly correlated with the economic value of its locality or neighborhood; and that, with its desireability in the eyes of the community.[9] The geographic distribution of the socio-genetic Strata thus constitutes a *geocline*.

Conclusions thus far

The Kingdom of Man is a single, inter-fertile species. This species displays six major psychogenetic Strata or sub-species, divided into at least twenty-one phenotypically distinct forms, a different set of forms for each Stratum in each of the 6 Periods in which it occurs. Each of these six Strata or sub-species is characterized by a series of ontogenetic stages or Sub-strata, numerically equal to the number of the Stratum in which they occur. Each Stratum is characterized by a genetically determined abstraction ceiling which no amount of training can exceed, but which is displayed phenotypically only through habitat-supplied training and education, at least equalling the abstraction ceiling in question. Each level of abstraction attained by an individual is expressed by the correct use of a distinctive vocabulary level and by the capacity to understand and control verbal behavior on all levels lower than its own.

The human psycho-genetic tree displays the distinctive form of a network or reticulum, Figure IV-2: the reason is that some descendents of older human sub-species regularly mutate into higher psycho-genetic sub-species. Such complete convergence happens rarely, if ever, in animals or plants. The reason is, of course, that we are discussing a *single* trait-complex, the one summed up under the heading *levels of abstraction,* but including levels of patience, tenacity, and the capacity to relate theory and practice to each other. Each Period's hierarchy of trait complexes is transmitted and altered genetically as a pleiotropic gene-complex. It is expressed phenotypically by a corresponding hierarchy of habitats (called social Strata), controlled by persons of an equal or higher psycho-genetic organization. Mutations occur in all Strata and result in upward and downward social mobility.

Human Strata behave like, and have to be classed as, true, highly dynamic sub-species. Their enormous mutation rates disrupt rigid caste systems, shattering the control structures of societies that do not admit their ever-emerging natural aristocracies to control positions. On the other hand, societies which do accomodate their ever-renewed natural aristocracies transmute themselves into higher and higher human Periods, each displaying a higher controlling socio-genetic Stratum.

These genetico-bio-psycho-socio-econo-political data arrange themselves as the Periodic Table of Human Cultures, whose framework is shown in Figure IV-4.

PERIODS			STRATA	GROUPS								
NUMBER	TOP STRATUM'S NAME	TOP STRATUM'S ACTIVITY	CHARACTERISTIC NUMBERS	I (0,−) ALLOPATHY	II (+,−) PARASITISM	III (+,0) COMMENSALISM	IV (+,+) SYMBIOSIS	V (0,+) ALLOTROPHY	VI (−,+) PREDATION	VII (−,0) AMENSALISM	VIII SYNNECROSIS	0 (0,0) NEUTRALITY
7	(PREDICTED) HIGHER INDUSTRIALISTS	Unified science, co-ordinated technology, controlled population, controlled pollution & resource depletion, public philosophy.	(PREDICTED) Seven Strata, 28 Substrata	→		TYPE SPECIMEN						
6	LOWER INDUSTRIALISTS	Separate scientific disciplines, uncoordinated technologies, uncontrolled population growth, pollution, resource depletion.	6; 6,7; 6 Omitted: 4 Strata, 14 Sub-strata 1 2 3 4 5 6 1,7 1,7 1,7 1,7 1,7 1,7 6 6 6 6 6 6	→		PARADIGMATIC CASES						
5	LITERATES	Writing, massive accumulation of information, great religions, theocracies, irreversibility of theory.	5; 5,7; 5 Omitted: 3 Strata, 9 Sub-strata 1 2 3 4 5 1,7 1,7 1,7 1,7 1,7 5 5 5 5 5	→		TYPE SPECIMENS						
4	HIGHER AGRICULTURALISTS, HIGHER PASTORALISTS	Irrigation, walled towns, large herding animals, priests, kings, nobles.	4; 4,7; 4 Omitted: 2 Strata, 5 Sub-strata 1 2 3 4 1,7 1,7 1,7 1,7 4 4 4 4	→		PARADIGMATIC CASES						
3	MIDDLE AGRICULTURALISTS, LOWER PASTORALISTS	Fertilizers, crop rotation, small herds, weaving, pottery, chants, shamanism, high chiefs.	3; 3,7; 3 Omitted: 1 Strat., 2 Sub-strata 1 2 3 1,7 1,7 1,7 3 3 3	→		TYPE SPECIMENS						
2	LOWER AGRICULTURALISTS, HIGHER HUNTERS	Slash-and-dibble gardening, use of hunting animals, chiefs, witch doctors.	2; 2,7; 2 1 2 1,7 1,7 2 2 [Key: Characteristic Number — Stratum; Sub Stratum; Kingdom, Group; Period]	→		PARADIGMATIC CASES						
1	LOWER HUNTERS GATHERERS	Hunting, gathering, Chants, ritual dances, Council, old men.	1; 1,7; 1	→		TYPE SPECIMENS						

PERIODIC TABLE OF HUMAN CULTURES

FIGURE IV-4 The Periodic Table of Human Cultures.

This framework is obtained by orienting Man's psychogenetic tree (Figure IV-2) upward, so that the progression proceeds from bottom to top, as in the preceding tables, and adding the nine Groups in their traditional positions. These Groups represent the nine possible coactions between any given culture's controlling Minority and its much bigger Majority. Following Ethel Albert, each culture's principal coaction is called its *dominant value-premise*, the others being deviant. It is this dominant coaction between work-component and controller which determines the major properties of any system, whether abiotic, biotic or cultural.

Two basic conclusions follow: There has never been, there is not now, and there never will be, a *classless* society.[11] And of all the possible coactions between psycho-genetic classes, the most creative and beneficent is the coaction displayed in Group IV: class *co-operation*, (+, +); the more the negative coactions occur, the more the culture disintegrates.

Since this Periodic table essentially restates what has been shown in Figures IV-1 through IV-3, further discussion of it does not seem necessary here. (See Appendix II at rear.) We turn therefore to a demonstration of its appicatlion in practical affairs.

Leibniz predicted (as already mentioned in Chapter II) that when his Universal Characteristic would be discovered then "If anyone were to disagree with me, I should say to him, 'Sir, let us calculate.' And, by taking to pen and ink, we would settle the question".[12] This table—which is little more than a traditional arrangement of these data, mapped into the Periodic coordinate system—is clearly a model of his General Characteristic: my students have used the coordinate system in what were once unresolvable political polemics. And we have found that, by taking to the blackboard, we could settle them as predicted. We therefore envisage an ever widening movement from the barricades to the blackboards; from the *multi*-versity to the new and higher *uni*-versity, and from break-down of our Period 6 culture to its transmutation into Period 7, Higher Industrialist. To see whether we now know how to go about it, consider the following mapping of a web-of-mind.

2. WEBS-OF-MIND: THEIR MAPPING AND SIMULATION[13]

A small web-of-life was mapped in Figure II-14. We progress now to map part of the similar, but higher and more subtle web-of-mind.

The first step is to map famous and familiar territory: the cybernetic structure of a stable and growing human ecosystem, translating its traditional terms into Unified Science's general and orderable terms, and setting our course by what Heisenberg calls our *compass*: our relationship with the central order.

This order, abstracted from the whole System-hierarchy, is System-theoretic in nature. It is thus, as Northrop advocates, deductively formulated and operationally verifiable. The second step will then be to map a previously bewildering detail within the bigger map. Namely, the exploratory "voyages" being made by some American colleges and universities whose destinations, this map and compass predict, will prove disastrous. By such prediction—and by computer simulation which this mapping permits and which, if opportunity to do so were provided, would be faster and cheaper—we might prevent irreversible destruction.

The big map represents the territory described by three explorers whose discoveries supplement each other and coincide with the central order described in Chapter II. Arnold Toynbee's *Study of History*,[7] Digby Baltzell's "*Protestant Establishment—Aristocracy and Caste in America*",[15] and Lloyd Warner and associates' Yankee City Series.[9] Using the characteristic numbers set forth in the Periodic Table of Human Cultures, the following map outlines the cybernetic structure of human cultures.

Its principles are precisely the same as those displayed by the preceding members of the Systems-hierarchy. (That is why we say that the universe is deeply simple.) Yet their human manifestation has important new characteristics. (That's why we say that the universe is richly strange.)

The strategic principle of all cybernetic systems is that the structure of the controller must correspond to the structure of the work component on one hand, and to the system's environment structure on the other, relating them to each other in such ways that the system survives.

Nobody, as a rule, disputes this statement. But when I point out the implication—*to survive, its controlling Minority must see the world as a Systems-hierarchy and think in terms of Unified Science*—someone usually retorts that no Minority in the world sees and thinks in these terms. This springs the trap: someone else then points out that this is true, and that consequently every industrial culture in the world is breaking down and disintegrating! If anyone objects, a few well known statistics clinch the point.

Periodic Table of Human Cultures

Would-be controllers and navigators are then ready to pay real attention to the new map. I admit that it is imperfect, but point out that with computer help it is perfectible. And what is the alternative? To go on muddling to the bitter end.[16]

This map represents the two main components of all lower Industrial nations, Period 6. It does so regardless of whether they are capitalist, socialist, communist, or fascist; and regardless whether they are Caucasoid, Mongoloid, or Negroid.

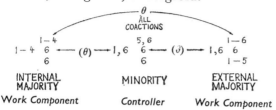

FIGURE IV-5 Work component and controller of human Cultures: Period 6, Lower Industrialists.

The center of this map represents this system's Minority or controller; its two outer parts represent the system's Majority or work component. According to the best data currently available (those in the Yankee City Series) the Minority is estimated to consist of the two highest Strata: Lower Upper and Upper Upper—about 3% of the population. The Internal Majority (on the left) is estimated to consist of the four lower Strata: Lower Lower (about 22%), Upper Lower (about 31%), Lower Middle (about 33%), and Upper Middle (about 10%). (This last is problematical: some of its members belong to the Minority.) The Majority (on the right) consists of human Periods 1 through 5; the highest of these, however, may contain six genotypic or potential Strata. For when their élites are educated in Lower Industrial countries, some of them sometimes display what appear to be Stratum 6 phenotypes. They belong, however (as yet) to the controllers of countries with fewer than six Strata—intellectual, social, and material—as these are defined in Figure IV-1.

This situation might appear anomalous to one-field specialists. Unified Science seems, however, to make it comprehensible. The following semantic glossary should make it more so:

Toynbee's famous terms *Minority* and *Majority* are obviously here in use. When their coactions are predominantly negative, he distinguishes the two kinds of Majority as *Internal Proletariat* and

External Proletariat. When their coactions with the Minority are positive, he still recognizes these categories, but calls the External Proletariat by relatively neutral names such as (for ancient Greece and Rome) *Barbarians.* Yet I think he would recognize and approve of the set I have called *External Majority.*[17]

Baltzell fully recognizes the Minority, which he calls the Establishment. He calls it *Aristocracy* when it is coacting positively, and *Caste* when coacting negatively with the Majority. He distinguishes both of the necessary kinds of cybernetic relations: the authoritative, which maintains the Minority's control; and the liberal-democratic, which keeps open its communication with the Majority. Baltzell also recognizes the Majority. He constantly shows important distinctions between the Inner Majority (those belonging to the same race and nationality as the Minority) and the External Majority (those belonging to nationalities and races different from Minority's). Though he does not name the latter as a set, I am persuaded that he would probably agree to the validity of this recognition.[18]

Continuing, now, to discuss our map, Lloyd Warner recognized the leadership and control function of the two highest Strata of Yankee City; also, several work and followership functions of the first four Strata. He showed that the control and leadership function had been unequivocal one and two generations before the study was made, as Baltzell shows it to have been throughout the United States—not as clear as it had been under America's Founding Fathers, but still indisputable. Warner showed two different changes going on simultaneously: *decline* of control itself, and *negativization of coaction* between controller and work component.

Control itself had deteriorated at the time Yankee City was studied: the public philosophy, as Walter Lippmann defined it, had widely broken down.[19] The Industrial Period was developing its own form of the old misery which Toynbee diagnosed in each of the disintegrating Literate civilizations and called *schizm of the soul.* (This condition can, as you will see, be corrected by the application of anthropology and Unified Science to Higher Industrial civilization, Period 7.[20]

"We are," George L. Williams points out, "trying to live in a scientific culture and enjoy the material benefits which applied science has showered down among us, yet not trouble ourselves deeply enough to become familiar with even the general root material of science. Few of us really want to know anything even approaching the whole truth about the world. Our external environment is modern; our minds are still medieval. As Mr. [H. G.] Wells wrote: '*Most of us prefer to float half-hidden even from ourselves, in a rich, warm, buoyant, juicy mass of familiar make-believe.*'"21

Controlling Strata hide this schizm of the soul from themselves because its recognition puts creative leadership out of the question, and they don't want to know it. They try to hide it from the Majority by means of the second change disclosed by Warner's study: they increase the social distance between them and the Majority. Thereby, however, they transform their aristocracy into caste, and change their coaction from creative leadership (+, +) into predation (−, +). With this, the society's "rhythm of disintegration"7 gets under way. See Yankee City, Volume IV, *The Strike*.22

The point is this: the breakdown of Yankee City's cybernetic structure—the decline of clear distinctions between work component and controller—is not a peculiarity of the United States, but a well known symptom of systemic breakdown generally. The basic and essential distinction between controller and work component had been clear under the nation's Founders;23 its breakdown had been foreseen by de Tocqueville early in the 19th century;24 and his prediction had reached a fairly advanced state of fulfilment when Warner and his associates studied Yankee City, and Baltzell studied America. I firmly believe Lloyd Warner would, if he were living, concur with this diagnosis.25

As for the External Majority, Warner and Lunt discussed a part of it in their chapter on "The Ethnic Minorities of Yankee City".9 Moreover, the entire Volume V of the Series is devoted to this part of the External Majority.26 Beyond this, however, Warner was a traditional anthropologist: he studied Period I peoples (Lower Hunters) and other sub-literate peoples in isolation from the Lower Industrial system, rather than as its External Majority. (He had to do this because these essential data were vanishing fast, and could never be retrieved, once they were gone!) He would though, I believe, have assented readily to this concept, which grows more obvious with each passing year.27

Formulation of the Effective University Social Structure.

When a complex mechanism has complexly deteriorated and broken down, the first thing the repair engineer needs to see is the blueprint of the system when it was new, and working correctly. For, as Abraham Lincoln said of the damaged social system whose controller he was trying to restore, "If we could first know where we are, and whither we are tending, we could better judge what to do, and how to do it." A breakdown can be made only relative to the condition of "normality." Figure IV-5 gives us this "base line" for Lower Industrial civilization, formulated deductively. Relative to this "blue print" of the normal Lower Industrial system, we can now map the structure of the "normal," "correctly" (cybernetically) working university.

The structure of the university's widespread current breakdown, actively engineered by social technologists who have misconstrued human kind's genetico-psycho-social structure, can then be understood much better. It is, of course, widely sensed without the benefit of maps. Sensing, however, is what the patient does. Curing is quite another thing: it begins with understanding of "where we are, and whither we are tending;" from there it sets its course.

The following map represents an ideal which administrations and trustees of top colleges in Britain, the United States and the U.S.S.R. have striven to approximate.

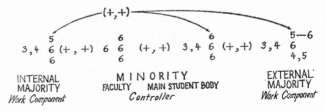

FIGURE IV-6 A web-of-mind: Academic institution for the transformation of top genotypic potentials into phenotypic actualities.

This university's objective is to train the highest possible proportion of the innately ablest young people for the most responsible positions, and to weld them into the effectively operating (cybernetic) controller of the nation's, and thereby to some extent the world's destiny. All of its *faculty* therefore belong to the Lower Industrial civilization (Period 6 at the base of their characteristic number) they all have the highest inborn (genotypic) potential (6 at the top).

The *main student body* (85%) have the same Period and Stratum as the faculty. Only their ontogenetic development stages are different: Sub-strata 3 and 4. This faculty develops these students optimally; and these students' consequent creative performance—especially when they have become part of society's controlling structure—raises faculty success, increases this faculty's reputation and rewards. Their coaction is thus strongly symbiotic $(+, +)$.

About 10% of the student body (left-hand entry) has the inborn abstraction ceiling of Stratum 5. But it has outstanding capabilities in other respects—say in various forms of art, athletics, or leadership. The roughly 5% of the student body from sub-Literate and Literate societies (right-hand entry) resemble the 10% in psycho-genetic ways, but differ from, and can thus contribute to, *all* of the other students ethnically.

Under the guidance of the Minority faculty and main student body, the Majority's students, Internal and External are strongly benefitted by them and by each other; and they in turn strongly benefit the two Minority groups. This is indicated by the five coaction symbols between these representatives of the Majority and Minority. This will now be demonstrated.

Discussion.

This map represents just the nucleus of education, and only during a civilization's genesis; its ectropic phase. What it omits is the control relations which develop outside academe. The following are examples of such developments: American, English, and Soviet Russian.

The first two relations are illustrated by a dinner conversation I once had in Geneva, Switzerland, in the Thirties with two English officials of the International Labor Office. We had been telling each other how students in our countries spend their summer vacations. *Their* vacations, it appeared, consisted of trips on the European Continent, largely among Upper Middle and Upper class people and in their surroundings. *Ours* consisted, by contrast, in any kind of work we could get, bar none. (I, for example, had worked as mess boy on two freighters, counselor in a childrens' camp, farm hand, tray washer in an employees' cafeteria, mountain guide in a resort, spool winder in a textile plant, and as attendant in a U.S. Veterans' hospital, predominantly for mental patients.)

The Englishmen were very pleased. They declared these summer jobs the most important part of my and my friends' education. This, they said, is the best way students can come to understand working people; their lives, their problems. They condemned their own kind of vacations as sheer waste, and held that in the long run it endangers the State.

K

130   Full Circle

Why these American students' summer jobs are important appears in the following little map of our school of hard knocks, the so-called "real" world. (Actually, of course, the academic world is just as real.)

The students learn the Majority's languages, idioms, and ways of thinking; enormously important things that even top-notch college professors rarely know, and which they could not teach, even if they did. The students learn the meaning of grinding physical work, of physical working conditions, of occupational hazards, brutality, hopelessness as no books, films, or courses can teach them. The foremen and workers among, and under, whom these students labor

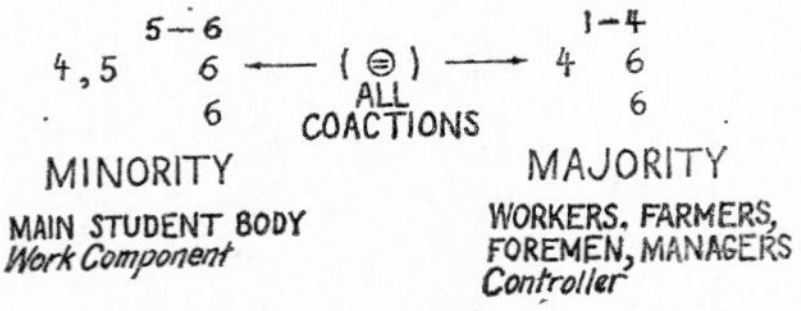

$$
\begin{array}{ccc}
 & 5\text{--}6 & 1\text{--}4 \\
4,5 & 6 \longleftarrow (\odot) \longrightarrow 4 & 6 \\
\cdot & 6 & \text{ALL} & 6 \\
 & & \text{COACTIONS} & \\
\text{MINORITY} & & \text{MAJORITY} \\
\text{MAIN STUDENT BODY} & & \text{WORKERS, FARMERS,} \\
\textit{Work Component} & & \text{FOREMEN, MANAGERS} \\
 & & \textit{Controller}
\end{array}
$$

FIGURE IV-7 A web-of-mind: Non-academic institution for the transformation of top genotypic potentials into phenotypic actualities. The "real" world: a school of hard knocks.

spot them and educate their characters in ways that can't be read; that have to be *experienced*. How to talk, to respond, to be respectful to people who expertly control plants, animals, things.—The university must be academic in order to think clearly and disinterestedly, as George Pake points out in "Whither United States Universities?"[14] Its education must therefore be constantly brought "down from the gallery of spectators and analysts into the arena with the contenders." This education is essential training for the exacting, hard-nosed task of governing a country. Many of America's Founders combined the two kinds of education mapped in Figures IV-6 and IV-7. As a result, they ran the country in a cybernetically feasible way.[23] Few modern American students, on the other hand, now get either kind of education. As a result "there has developed in this country," as Walter Lipmann pointed out long ago, "a functional derangement of the relationship between the mass of the people and the government".[19] This derangement will presently be specified and mapped.

But first, a report on the combination of these two complementary aspects of a Minority's education by the Soviet Government. This report was made in 1967 to the New York Chapter of the Society for General Research by a well known personage who had just returned from an extremely friendly visit to the U.S.S.R. The meeting was held at the New York Academy of Sciences under my chairmanship, and I paraphrase my notes.

There are in the U.S.S.R. certain unpublicized schools whose object it is to train the country's most responsible future leaders and controllers. The most outstanding children are sought out by means of the best available tests and observations. (The youngest ones enter the school when about eighteen months old.) Absolutely no expense or effort is spared.

At first the ratio of children to teachers is 1:1; later it becomes 2:1 and higher. Every effort is made to assemble the kind of faculty and students mapped in the center of Figure IV-6. As these students grow older, individuals are sent periodically to work in carefully chosen factories, collective farms, research centers, army units, and so forth. This gives them the training mapped in Figure IV-7. (Soviet administrators refer to these superlative institutions—half-jokingly, of course—as their schools for "philosopher kings.") From time to time individual students take courses in various gymnasia and universities open to all who pass public examinations. This serves to give the future leaders connections in, and understanding of such institutions.

This is as far as I paraphrase my notes. It should, however, be pointed out parenthetically that these students' sporadic attendance at public institutions also serves the important function of veiling their strategic schools' existence. This veiling is not arbitrary: the Majority people—as Figures IV-2, IV-3, and IV-4 indicate—having lower abstraction ceilings, are more or less aphasic and agnosic to the Minority's highest and most important levels of abstraction. They do not clearly understand the culture's need for such exclusive schools. So, if they knew of these schools' existence, they would demand open admission to them, breaking them down to the Majority's lower ceilings, as diagrammed below. This would, in time, bring to the Soviet Union what Lippmann calls the "functional derangement between the mass of the people [Majority] and the government [Minority]."

To prevent this, and yet maintain such special schooling openly, the authorities would be obliged to explain and massively advocate the cybernetically effective relation between Majority and Minority. But this would seriously hamper their currently successful disorganization of education abroad, not to mention openly contradicting their promise of "classless" society some day in the Soviet Union.

Both of these unacceptable alternatives have been avoided by keeping secret the existence of Soviet schools for Philosopher Kings.

How, then, could the public disclosure of them in the present book be countered? Two ways present themselves: these schools' existence could be denied, and this description be denounced as fabrication. That would however be quite useless since, as the authorities know, this disclosure can be backed by much more detailed evidence, including first hand accounts.

The other alternative would be to present these extraordinary institutions as a great Soviet achievement, and as an advance toward the endlessly promised "classless" society.

How could this claim be supported?—By showing, with detailed evidence, that the students are not drawn exclusively from the "New Class," the Soviet Minority (the controlling part of its hierarchy), but that they come from the entire population and are chosen on the basis of fair and objective tests of ability and character.

Whether this is in fact the case could, of course, be established only by an investigation of the Stratum position of every student and alumnus, and by analysis of the Soviet tests. (The shoe would then be on the other foot, putting Dr. Jensen's critics in a position which would help separate demagogic agitators from serious and fearless scientists.)

To clarify the issue, let us make the improbable assumption that the Soviet authorities made this second claim and that, with their full cooperation, it was demonstrated by outside investigators to be accurate. The first conclusion would then be that the Soviet Union is moving toward realization of Thomas Jefferson's model of government: government by what he called the *natural aristocracy*. "I agree," he wrote to John Adams, "that there is a natural aristocracy among men. The grounds of this are virtue and talents . . . The natural aristocracy I consider as the most precious gift of nature for the instruction, the trusts, and the government of society."[28] (If Communists were now to claim that by *classless society* they mean government by the natural aristocracy we would, by taking to pen and ink, have started to settle the question.)

The second conclusion would be the one predicted by Unified Science: The fairest and most objective tests will show virtue to be quite evenly distributed among all social Strata. But they will show a great skewing of talent-distribution, where *talent* is predominantly identified with abstraction ceilings. (This skewing occurs both empirically and *ex hypothesi*: in Unified Science the distribution of social Strata is based upon the distribution of abstraction ceilings, as shown in Figures IV-2 and IV-4.)

The third conclusion follows inexorably: the use of fair and objective tests results in a vast preponderance of high Stratum (Minority) students in those schools and universities which are designed to train young people for "the government of society." It is therefore natural that the Stratum representation in such schools should approximate the ones shown in Figures IV-6 and IV-7, beginning to settle that question too.[29]

The Opposite Derangements of Government.

Our model of Leibniz's Universal Characteristic can now be applied to rational settlement of the major reciprocal political questions of our century: the breakdown of working democracies, on one hand by egalitarian democracy, and on the other hand by dictatorial autocracy. Walter Lippmann has analysed what has been happening quite effectively; that is to say, cybernetically. A few short excerpts from his famous book, with the appropriate interpolations, will complete the large and dynamic map of Industrial civilization outlined above. The consequences of open (blind) university admission policy on one hand and of the development of well matched key-lock colleges, universities, trade-schools etc. on the other hand will then become much clearer than before, making prediction possible. We will conclude by verifying our prediction in terms of a concrete case: New York's City College.

"The more I have brooded upon the events which I have lived through myself," Lippmann wrote in 1955, "the more astounding and significant does it seem that the decline of power and influence and self-confidence of Western democracies has been so steep and so sudden. [It has been steeper yet in the years since he wrote this presage.] We have fallen far in a short space of time... What we have seen is not only decay—though much of the old structure was dissolving—but something which can be called an historic catastrophe."[19, 30] p.15.

And what is this catastrophe? "In the effort to understand the malady of democratic government," Lippmann replies some forty pages later on, "I have dwelt upon the underlying duality of functions: *governing*, that is, the administration of the laws [which cyberneticians call *control*] and *representing* the living persons who are governed, who must pay, who must work and, it may be, die for the acts of the government [which we call *feedback* to the controller from the output of the *work component*]. I attribute the democratic disaster of the twentieth century to the derangement of these primary [cybernetic] functions.

"The power of the executive [the controller] has been enfeebled, often to the verge of impotence, by the pressures of the representative assembly and of mass opinions [the work component]. This derangement of the governing power has forced the democratic states to commit disastrous and, it could be, fatal mistakes. It has also transformed the assemblies in most, perhaps not in all, democratic states from defenders of local and personal rights [of the two components] into boss-ridden oligarchies [lower level, predatory controllers], threatening the security, the solvency, and the liberties of the state [the social system as a whole]."[19] pp. 54–55.

Lippmann now states the systems-theoretic meaning of democratic constitutions: "In the traditions of Western society, civilized government is founded on the assumption that the two powers [controller and work component] exercising the two [cybernetic] functions will be in balance—that they will check, restrain, compensate, complement, inform and vitalize each other."

His diagnosis, stated in the same non-scientific language is, that "In this century, the balance of the two powers has been seriously upset. Two great streams of evolution have converged upon the modern democracies to devitalize, to enfeeble, and to eviscerate the executive [control] power."

There is another stream which Lippmann discusses later on, but which precedes and underlies the ones he describes here: the stream of specializations which has eroded the West's awareness of the cybernetic nature and operation of their psycho-political systems, making its statesmen and political scientists helpless to understand the catastrophe, and prone to increase it by their very efforts at correction.

"One [great stream]," Lippmann continues, "is the enormous expansion of public expenditure, chiefly for war and reconstruction; this has augmented the power of the assemblies [representing the work component] which vote the appropriations on which the executive depends. The other development which has acted to enfeeble the executive power [controller] is the growing incapacity of the large majority of the democratic peoples to believe in intangible realities. This has stripped the government of that imponderable authority which is derived from traditional, immemorial usage, consecration, veneration, prescription, prestige, heredity, hierarchy."[19] pp. 54–56.

The plight of the governments of what was once the Creative Center (Figure II-16b) is herewith affirmed to consist significantly in the disorganization of the controller (the executive branch), and the unorganized, chaotic over-growth of the work component (the electorate and its representatives). Exactly the converse, Lippmann affirms, is the defect of the totalitarian ideologists: the governments of Extreme Left and Extreme Right, mapped in the same Figure.

"In the reaction against the practical failure of the democratic states," he goes on, "we find always that the electoral process is shut down . . . and that the executive function is taken over—more often than not with popular assent—by men with a special training and a special personal commitment to the business of ruling the state. In the enfeebled democracies the politicians have with rare exceptions been men without sure tenure of office. Many of the most important are novices, improvisers, and amateurs. After a counter-revolution has brought them down, their successors are almost certain to be the elite of the new revolutionary party, or an elite drawn from predemocratic institutions like the army, the church, and the bureaucracy. . . The post-democratic rulers are men set apart from the masses of the people. They are not set apart only because they have [and use] the power to arrest others and to shoot them. They would not long hold on to that kind of power [by itself]. They have also an aura of majesty, which causes them to be obeyed. That aura emanates from the popular belief [and often the fact] that they have subjected themselves to a code and are under a discipline by which they are dedicated to [and seem able to *achieve*] ends that transcend their personal desires and their own private lives" pp.59–60.[19]

These insights, being correct, permit us to complete and sharpen Lippmann's diagnosis: the breakdown of the democracies—their executives' loss of self-confidence, authority and power, and their electorates' loss of equally important power to believe in and follow the executives—is due to the break-up of what Lippman calls *the public philosophy*, and what poets more powerfully call the *Circle*.

"The Circle of Perfection," says Marjorie Nicholson, "from which men had long deduced their metaphysics, ethics and aesthetics, was broken during the seventeenth century. 'Correspondence' between macrocosm [the universe] and microcosm [the human mind], which man had accepted as basic to faith, was no longer valid in [his incorrect and unintegrated picture of] a new mechanical universe and mechanical world."[31]

The unity of mind and feeling, the *Circle of Perfection* created by medieval mystics and scholastics, was shattered by the apostle of empiricism, Francis Bacon, and the parts were dispersed by his fellow empiricists, each of whose strong but little knowledge became ever more clearly a deadly dangerous thing.

"To Bacon [and to empirical one-field specialists ever since]," Nicholson goes on, "the Circle of Perfection was no more than a 'fiction', and the tendency of man to find it everywhere on earth and in the heavens one more indication of the dangerous haziness of thinking he called an 'Idol of the Tribe': 'The human understanding', he said, 'is of its own nature prone to suppose the existence of more order and regularity in the world than it finds'".[31] Thus, being human, Bacon displayed and worshiped the idol of his own new Tribe: the tribe of empirical one-field scientists who, by their nature, see individual parts of the universe, but not their structural correspondence and assembly.

Sir Francis Bacon and his disciples were correct in shattering the spurious seventeenth century Circle, for many of the parts comprising that assembly of knowledge have proved demonstrably wrong. The scientists were also correct in patiently and tenaciously discovering the future Circle's empirically valid component parts, one by one, in spite of the deadly, excruciating irrelevance and meaninglessness of the resulting storehouse of unassembled parts, the multiversity.

In overcoming the dangerous medieval haziness of thinking—in carrying out the 18th century's response to that challenge, fatuously called the Enlightenment—these scientists created the twentieth century's far more dangerous challenge: our deep *Ensombrement,* which Lippmann calls *The Eclipse of the Public Philosophy*[19] (Chapter VIII). This was not due to viciousness or stupidity, but to the structure of evolution, inherent in the Systems-hierarchy, which poets have called *the darkness before the dawn.*[32]

As late as the eighteenth century, men held what Lippmann calls the doctrine of natural law: the certainty that there is law "above the ruler and the sovereign people . . . above the whole community of mortals."[33] But in the twentieth century's Ensombrement there is "a plurality of incompatible faiths"[34] and multitudes of agnostic, existentialistic and nihilistic people. And far worse yet, there are millions of ideologists in the totalitarian countries and infiltrated throughout the egalitarian democratic states for whom strong

Circles of Perfection had been closed prematurely, and incorrectly, back in the 19th century: namely, Marxists and fascists.

Because of this profound Ensombrement, Lippmann points out,

> The democracies of the West became the first great society to treat as a private concern the formative beliefs that shape the character of its citizens.
>
> This has brought a radical change in the meaning of freedom. Originally it was founded on the postulate that there was a universal order on which all reasonable men were agreed: within that agreement of the fundamentals and on the ultimates, it was safe to permit, and it would be desireable to encourage, dissent and dispute. But with the disappearance of the public philosophy and of a consensus on the first and last things—there was opened up a great vacuum in the public mind, yawning to be filled.[19] p. 100.

One thing that can help fill this vacuum is the late twentieth century's Circle of Perfectibility: Unified Science.[35] The one-field sciences have now come full circle: their assembly *consists* of these powerful sciences and technologies; it *displays* the positive value bias of the most deeply sensed religions; and it *corrects* the Marxist misinterpretation of history and the fascists' and liberals' opposite misinterpretations of genetics and education.

The parts of Unified Science are mutually illuminating, reciprocally correcting, and Circle expanding. Among these parts is the microcosm: the mentalities of all the human Periods and Strata, from naked Hunters and Gatherers in dry Australian deserts to space-suited, computer-guided astronauts on the airless moon. Unified Science conveys to each of these microcosms, and internalizes within it, the macrocosm's verifiable and compelling natural and moral law.[34]

In the light of this Full Circle, the above-mapped 19th century American and 20th century Soviet webs-of-mind gain meaning and permit prediction. For in both of these cultures the distinction between controller and work component, Minority and Majority, is clear, and their ordination is cybernetically correct. Each of these systems had, or has, a public philosophy at the time in question which accords with the central order in regard to the superordination of controller over work component. Unified Science permits us to formulate the problem of open and selective admissions and to predict the outcomes of these alternative solutions as follows:

Equalization of Opportunity in Our Stratified Population.

All civilizations equalize the education opportunities they provide for their various age-grades or Sub-strata, from infancy to maturity, by corresponding gradings of the subject matter taught. Equality of opportunity for unequal age-groups is approached by careful provision of a corresponding inequality of educational habitats.

So also in regard to their various psycho-social Strata: equality of opportunity for people with unequal inborn abstraction ceilings is approached by careful provision of corresponding training situations. During the first few years (preceding and up to the attainment of the first human abstraction ceiling) a single institution, Kindergarten, and the first few grades, provides equality of opportunity for all Strata.[36] Figure IV-3 shows graphically humankind's initial state of intellectual identity (zero), and its divergence into Strata during ontogeny. As the childrens' creodes separate—as Stratum after Stratum approaches its abstraction ceiling, levels out, and is surpassed by the people with higher inborn abstraction ceilings—each Stratum enters the corresponding set of educational institutions. Namely, the kind of institution designed to provide for it the opportunity to realize its inborn capacities to the fullest degree: apprenticeships, craft schools, trade schools, secretarial schools, high schools, preparatory schools, (European) gymnasia, junior colleges, colleges, institutes of technology, graduate schools, post-doctoral training courses, institutes for advanced study, and so forth.

In Figure IV-2, Period 6 (Lower Industrialists) displays six Strata, each characterized by the corresponding number of Sub-strata. The highest Sub-stratum in each case (including the first one) is reached by, and only by, utilizing opportunities for continuous, persistent development of inborn capabilities.

Exceptions occur for reasons well known to geneticists, and are often important. Child prodigies appear from time to time, for whom equality of opportunity requires skipping one age-graded school class after another; for instance, Norbert Wiener.[37] Stratum prodigies occur, for whom equality of opportunity requires the by-passing of one school-type after another; John Stewart Mill.[38] Period prodigies occur, for whom equality of opportunity requires travelling to an Industrial country and studying in its higher schools; for instance, Yomo Kenyatta.[39] (These latter two kinds of exception comprise the two Majority groups mapped in Figure IV-6.)

Downward exceptions also occur, especially in the highest and most recently entered Strata. (Geneticists recognize them as "regressions toward the mean."[40]) But downward exceptions occur in all Strata. (Of late those who display them have been euphemistically called "retarded.") For them, equality of opportunity requires repetition of school classes, top Stratum children's apprenticeship in trade or craft schools, and emigration to less developed regions or countries.

Since respectable institutional channels for these adjustments are inadequate or absent, considerable numbers of young people, called *Hippies*, are improvising equality of opportunity by "dropping out" in all three of these ways. Many emigrate to rural parts of New Mexico, Arizona and similar regions and try unsuccessfully to simulate pre-Industrial life styles.

The term *dropping out*, however, is also used to denote the opposite non-institutional adjustment. Where formal institutions have broken down and deteriorated—whether actually or apparently—some among their employees who recognize this fact refuse to deteriorate with them, and "drop out," sometimes at great financial sacrifice.—See for example, "Dropping Out in Manhattan" by Colette Dowling.[42]

This kind of withdrawal has been going on for millennia. The most significant personalities in history, as Arnold Toynbee shows abundantly, have withdrawn ("dropped out") and then, *having integrated their minds and personalities*, returned and either halted their cultures' downward retreats or contributed to their eventual upward transmutation into higher Periods.[43]

One of the most important weapons for changing this rout into an advance is a battery of tests for accurately diagnosing each individual's genetic capabilities (keys) and matching them with the corresponding educational and professional habitats (locks). Among the most strategic people of our time, therefore, are the psychosocial scientists who are developing such "key"-testing methods. Namely, aptitude tests, intelligence tests, personality measurements, and so forth, as described in this chapter's second part by Arthur Jensen. These efforts must be assembled and then extended to the corresponding "locks": to the educational and professional institutions in which these individuals develop their capabilities and then employ them, as outlined in Figures IV-6 and IV-7. We have the necessary components of Period 7 ecosystems. What we now need, in order to assemble them, is practical methods for spotting and finding these many different parts, and putting them together into a viable whole. Success in doing this would make our culture itself "drop out" in the *upward* sense: if we succeed our culture will drop out of the disintegration Groups of Period 6, and up into Period 7 (Figure IV-4); if not, it will drop *down*.

What is the main obstacle to this upward development? It is the ideologies of Left and Right, described in Chapter II, whose confrontations our universities and scientific societies—warehouses of unassembled disciplines—are ever less able to resist effectively.—See, for instance, "Whither United States Universities" by George E. Pake.[14]

Under persistent attacks of Left-Center liberals—whose genetically ignorant discipline Baltzell calls the Social Gospel and the New Social Science[15]—under the vehement attacks of Far Left, and violent attacks of Extreme Left radicals, the National Academy of Science formally renounced the development and application of intelligence, aptitude and personality tests for a whole year! Being severely hampered by one-field specialization in distinguishing pseudo-science from science in this field, its subsequent reversal of this tangibly anti-scientific position still remains more apparent than real.

Unified Science, however, provides a clear, verifiable hypothesis and model for correcting this fatal defect. According to the theory presented in Chapter II, a strategic aspect of our culture continues to loiter on the threshold of Period 7, Higher Industrialists. Namely the genetico-psycho-socio-econo-political sciences and technologies. This failure to consolidate social science creates in our culture's controller or Minority what Lippmann called "the great vacuum yawning to be filled." That is to say, a lack of coherent understanding, *resulting in persistent failures of leadership*. Specifically it preserves the Minority's incapacity to diagnose malfunctions, prescribe feasible solutions, prognose the outcomes of alternative prescriptions, and then execute the most promising one effectively.

Into this leadership vacuum rush the ideologists—people such as the totalitarian democrats and various fascistic racists, whose worldviews were prematurely unified in the nineteenth century; unified before the rise of modern physics, chemistry, biology, genetics, or any other modern science; and by non-scientists at that. Their misinterpretations of history, genetics, psychology, and so forth are, however, systematic and mutually reinforcing. This gives them the confidence which our traditional leaders lack, and therewith the power to mislead the Majority disastrously.[44]

How have we found out that they are misleading our education and our culture? In the same way that physical and biological scientists find out when they are misled: by making theoretical models and subjecting them to experimental verification. "The verification of a model such as occurred with Rutherford's nuclear atom can greatly extend the range and scope of the physicist's understanding," say physicists Kendall and Panofsky. "It is through the interplay of observation, prediction, and comparison that the laws of nature are slowly clarified."[45]

Our maps of alternative webs-of-mind are models of genetico-psycho-social systems. Figures IV-6 and IV-8 predict the coactions to be expected in universities containing certain proportions of Minority students and Majority students, Internal and External. These predictions can now be compared with observations of the two corresponding kinds of institutions.

Figure IV-8 is a theoretical model of an open-admissions college, as this term is defined and applied at New York City College, 1970–1971. It will be compared with the earliest available report of the empirical event: "Up the Down Campus—Notes from a Teacher on Open Admissions" by M. Ann Petrie,[46] a member of the College's English Department who strongly supports this so-called "open" structure.

The only screening used in the admission of new students was a school average of 80 or top-half rating in high school. According to Miss Petrie "850 of the 2,440 freshmen who registered as full-time day students last fall would not have been there had traditional standards been applied." This increases the ratio of Majority to Minority students in the symbiotic model (Figure IV-6) by about 35%, making their proportions roughly equal (50%–50%). According to coaction theory, a model displaying these changed proportions predicts the following change of coactions.

This model assumes the same faculty, and the same 10% of exceptional, top-capability Internal Majority and 5% of top capability External Majority students as displayed in Figure IV-6. It maintains this grouping in contradiction to the empirical phenomenon with which it is compared: Ann Petrie quotes Alan Fielin, dean of the open admissions program as saying that "Our freshmen do not arrive as two, homogeneous packages—'regular freshmen' (Minority Students) and 'open admissions freshmen'. . ." To these "regular freshmen" (Minority Students) it adds roughly 25% of Internal and 10% of External Majority students who "would

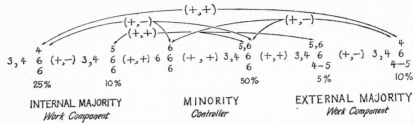

FIGURE IV-8 Open admissions institution for the ostensible "transformation" of medium genetic potentials into top phenotypic actualities. A prediction model.

not have been there had traditional standards been applied," decreasing the proportion of Minority students by the corresponding 35%. (This corresponds quite accurately with Petrie's report.) "At midterm, the dropout rate, less than 11 per cent for City University as a whole, was no worse than it had been in the two previous years. But because of the commitment to keep the open admission students at least three semesters regardless of grades, that's scarcely illuminating." Traditional standards, then, remained unapplied to these 35% of Majority students after, as well as before, open admissions, maintaining roughly these proportions throughout the academic year in question.[62]

Our theoretical model—based as it is upon the massive data organized in Figures IV-1 through IV-4—predicts the following: In contrast to the Exceptional Majority freshmen, the great majority of the open admission students will rise somewhat above their home habitat in regard to vocabulary level (Figure IV-3). But they will not move to a higher social Stratum. Instead, they will presently revert to little above their previous level of reading, speaking, and writing skills, just as do most of those who, through special training, rise to a higher vocabulary level (Figure IV-3). However, having been left untrained for occupations they could have filled successfully if they had been given *actual* equal opportunity—that is, if they had been accurately tested and given a chance to fit themselves into the "locks" for which they have the inborn turning capability—they will find themselves in, and put their community in, a desperate position; one which will threaten to wreck the community and them together.

This situation has already begun to appear inside City College "Some faculty members. . ." says Petrie, "perceive the open admissions students not merely as different from the stereotyped City student, but as inferior to him, and they hate the changes in curriculum and classroom styles that may have come because of the difference." One such professor told a student that "he didn't have time to waste on problems a ninth grader should know the answer to."

What are these changes in curriculum and classroom styles? They involve a lowering of abstraction ceilings and of vocabulary levels; *a lowering of the essential training for which the College was established, and which every Industrial civilization's Minority has got to have if their culture is to survive, let alone to advance.*

That these changes are hated is natural. And they are hated not just by professors, but by the qualified students of both Minority and Majority, whom they *prevent* from receiving the high-level training for which they have the inborn capacity, and which they must get if they are to fill the leadership vacuum, correct the system's cybernetic derangement, and halt our society's catastrophic decline.[47]

These "changes in curriculum and classroom style"—this decline in the level of thought and education—will rapidly change this institution's reputation in the same direction. Its academic degrees will decline in competitive value, and the coactions between the majority of open admissions students (mapped at the outer ends of this web-of-mind), and the rest of the students, the faculty, and the community *will consequently become objectively negative*. They will become so, regardless of what they appear to be subjectively to those who gladly sacrifice time, effort, money and reputation in the hope of benefiting the slightly above average Majority, but who in reality damage far more of them than they help.[48]

Petrie reports a change in objective coactions between those who pass the tests, and most of those who are admitted without it. Since the latter remain, regardless of whether they do the work successfully or not, the former are leaving City College in considerable numbers. If the students with lower potential benefitted, this coaction would be: benefit to the weak, damage to the strong $(+, -)$, which is bad enough. But if both kinds of students are damaged, as our model predicts that they are being damaged, their coaction is mutual harm, synnecrosis $(-, -)$; social disaster. Among each group's members, however, the coaction will probably be some degree of symbiosis $(+, +)$. That will polarize the community and prepare civil war.

"I would be very surprised," said Dean Fielin, "if our open admissions experience did not result in changed views about education for all freshmen, indeed all students."[46] This model predicts that, if accurate records are kept on all these students for the next ten years, the greatest surprise, and the view that will change the most, will be Dean Fielin's own.

Conclusion.
Unified Science's unequivocal prediction is that open admission not only deprives the Majority of the education that could advance it most, but robs the Minority of equal opportunity to realize its inborn potential, and the community of "nature's gift for the instruction, the trust, and the government of society." These predictions should accelerate the development, expansion, testing and application of *culture-fair diagnostic tests* for individuals ("keys") on one hand; and on the other hand, of complementary tests for all kinds and levels of instruction ("locks").

3. FROM MULTIVERSITY TO UNIVERSITY

These two developments, however, subsume and pre-require the third: assembly of the Lower Industrialists' self-destroying mental and spiritual chaos of unassembled and amoral disciplines into the Higher Industrial Circle of Perfectibility.

"I place science within the area of accumulative knowledge," declared James B. Conant in his 1947 address as retiring President of the American Association for the Advancement of Science, *and urged the coordination of its role in our society.*[49] This address opened the door to the sustained effort whose results are here introduced. The uncoordinated accumulation of scientific knowledge in our traditional universities which has brought on the derangement and decline of Western democracy is diagrammed in Figures IV-9 and 10. The result of coordinating this accumulation into the coherent world understanding essential to successful leadership of society will be shown in Figures IV-11 and 12.

At its left, this figure represents the four traditional divisions of the university—those, for instance, which participated in the Interdivisional Committee to whose Chairman this book is dedicated. One of them, the Division of Biological Sciences, is shown in some departmental detail.

Two biology departments, being markedly different from the rest, are underlined: ecology and paleontology. They deal with whole systems of the most important kind, natural empires. They are outstanding because, as Ulrich Sonnemann points out, "A whole [system], whether encountered by the physicist or the social scientist

Periodic Table of Human Cultures

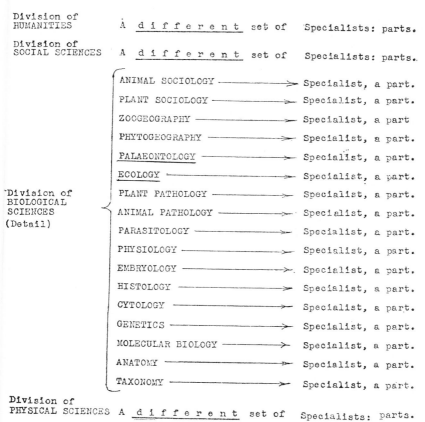

Figure IV-9 The Multiversity—a parts-making industry with no assembly plant.

[or the ecologist], is a lawful context which gives significance to each particular part-phenomenon that articulates itself within it; what makes its inner lawfulness understandable, however, is its own overall significance which it derives from the broader [systemic] context in which it is embedded and to which it refers." Sonnemann, who is writing on *The Specialist as a Psychological Problem*, then goes on to anticipate our discussion of what may be called the *organized specialist* and the *generalist:* "The true academician's [the generalist's] subject matter, *in principle,* becomes the universe: whatever he encounters—and it may lie in exceedingly small sectors of fields—occurs to him in such ways as to represent a universal order. Such a universal order, since it already determines the phenomenal structure under the scientist's observation, is inseparable from the structure."

Sonnemann then characterizes the rest of the departments shown and implied in Figures IV-9 and 10 as follows: "To the extent, then, to which it [the whole system] drops out of sight, to the extent to which his [the specialist's] preconceived procedure interferes with the self-articulation of any subject under his attack, phenomenal structure will escape, first his eye, ultimately his theories."[50] These people's departments deal with sections of systems. And, as Sonnemann points out, "Implicit in all sectional science, the arbitrariness of primary determination of subject matters which of their own natures are universes [systems] does not, apparently, make a science any more analytical; the typical specialistic approach . . . is characterized at least as much by his blindness for relevant detail as for wholes."

He then sums up as follows: "The loss of the criterion of intrinsic truth [namely, the systems criterion] is inextricably linked with the sectional character of the [non-systematic] sciences themselves. . ."[50]

At the right, our figure represents the multiversity's product: specialists. That is to say, mental parts which have not been designed for assembly and for which it has no assembly plant. You will, of course, note two exceptions: the products of the departments of ecology and paleontology, who are incipient generalists. They are incipient because they are unassemblable parts; nonetheless, the systems they study do encompass all their colleagues' subject matters. Yet they have no way of assembling them, no effective assembly plant or technique. And their specialized colleagues, for reasons shortly to appear, have no way of grasping the ecologists' (as also the anthropologists', historians', and atomic physicists') basic differentness. So they see and treat these near generalists simply as somewhat special or peculiar colleagues, and go on as before.

The profound difference between these two groups—between what Sonnemann calls legitimate and illegitimate specialisms—will shortly be diagrammed. "No wrong attaches to any specialization, any concentration on a particular subject matter, or realm of subject matters," he points out, "which, in setting its method of analytic attack, closely follows the given structure of the subject.

If the subject happens to be a whole [system], such as the whole subject matter of entomology, or of its self-articulated subdivisions [which is ecology], the wholeness of the subject implies at once the *universality* of good order constituent of nature throughout, and a distinct *separateness* from other subject matters of nature: a separateness which, in favoring the concentration of the scientists' focus upon it, legitimizes, at the same time, its specialistic restriction.

"It is different," he points out, "for such subject matters of one science as are inseparable from other subject matters lying within a different science which in actuality form one [system] with them. Specialistic narrowing of focus here cannot but fail to perceive the order of the whole and cannot help replacing what it misses by a mechanical order which it imposes on the subject by means of procedure." He then goes on to point out:

> Bergson has already stated the [illegitimate] specialist's inclination to conceive of his subject matter in terms of his method rather than the other way around. [But] he had not pointed out in any detail . . . the various slights of hand [read *unconscious mistakes*] which turn abstractions from processes into factorial [empirical] entities assumed to partake in the process as such. . . Where the statistical method is used, not for its legitimate end of clarifying the structure of large bodies of relevant data but of predicting the structure of human events to come, [illegitimate] specialistic attitudes are characterized by an inclination to reduce *events* to mere *occurrences*, focusing on their comparative numbers without questioning the basis of comparison used, remaining blind to the specific event-nature of each, and never inquiring into those elemental processes behind them—revelation of which would invalidate the cherished technique from the start.[51]

He then touches upon the heart of the danger in which illegitimate specialists involve our culture; the self-perpetuating cybernetic structure displayed in Figure IV-10, which Mr. Stafford Beer has called the *meta-threat:* "The circularity of the method, a closed system inaccessible to any such observations as would interfere with its own premises. . ." The specialist's conscious input comes from his own non-systemic field. His detector screens out inputs from the rest of the system of which he is studying a part; and from other functionally related ones. His conscious (subjective) output goes to his own traditional field, though his objective outputs may go disastrously to others as what he calls "side effects." These are, however,

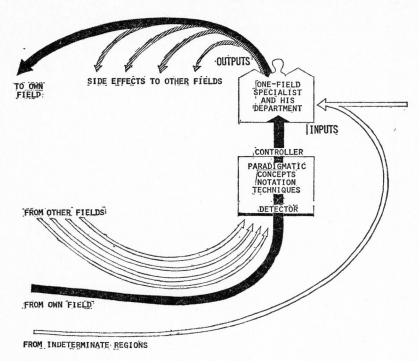

FIGURE IV-10 The unorganized specialist—unassemblable product of the multiversity.

screened out by his detector, preventing him from becoming aware of and correcting his mistakes; also, from listening to or understanding those who try to point them out. We shall return to these figures in Chapter V.

This is the suicidal structure of the multiversity's major output, Figure IV-9, most of its graduates. Failure to analyze its own self system-theoretically; failure, in fact, to develop what Sonnemann calls "the criterion of intrinsic truth" results in "the sectional character of the [non-systemic] sciences themselves."[50]

What is to be expected of such a fundamentally mis-designed system as the multiversity? It is structured to commit what the Bible calls the *Unforgivable Sin:* the kind of sin of which one cannot become aware, which therefore cannot be repented or corrected. The multiversity's expectable output is a cumulative series of disasters. And it is living up to expectations.

"The problem we now confront," says George C. Lodge in *Harvard Today*—"pollution, population, transportation, education, even perhaps malaise on the assembly line, and alienation and distrust—are not soluble pragmatically." And he puts his finger on the source of trouble: "The traditional sense of the word 'science' implies specialization. The transformation in which we find ourselves is 'unscientific' in its assertion that specialization, concentration on the parts, is not the way to a useful whole."[52]

By confining each of our millions of students to one or two of these artificially segregated, meaningless disciplines, with its special vocabulary and background theory, the multiversity produces on one hand what Jere Clark calls the ignorance explosion, and on the other elimination of morality and wisdom. As Lodge puts it, "Paternalism won't work [any more] because there is no father and there are no children."[52] This implies the collapse of leadership; the catastrophic derangements of both democratic and dictatorial forms of government, and the visible danger to the survival of Man on Earth.

The New University.

Stafford Beer affirms that "The fresh design of a meta-system, exerting meta-controls, is the *only* solution to our problem. The problem is for cybernetics to discover, and to make abundantly clear to the world, what meta-systems truly are..." In fact, he says, "We should create a meta-system to handle the meta-threat."[53,54]

Our Council has been working hard at this for twenty-odd years, and I a good deal longer. Let's first look at the over-all structure of the New University which we propose; and at its strategic products, the generalist and the organized specialist. Then we can analyze its internal structure, the meta-theory and meta-language which make these products possible.

This is the proposed blue print of the New University; of its structure over-all, Figure IV-11.

In the lower left-hand corner, C. P. Snow's Two Cultures, Literary and Scientific, are shown modified and assembled into a new form of what Walter Lippmann has called The Public Philosophy.[41] Namely, the Public Philosophy of our new, emergent culture, Higher Industrial Civilization; Human Period 7 of the Periodic Table of Human Cultures, Figure IV-4.[55] This Public Philosophy emerges as a result of certain quite precise modifications of

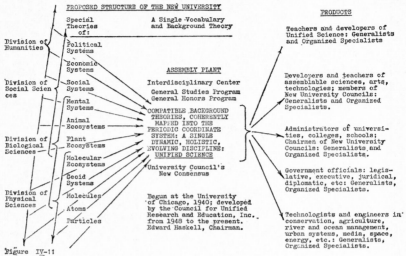

FIGURE IV-11 The New University—structure and product.

our traditional scientific and literary cultures, as also of the conflict-biased Marxist theory and practice, diagrammed in *Plain Truth—And Redirection of the Cold War*.[23]

These changes are already in process of occurring: first because of the new concepts and methods of Unified Science, represented by the three nested braces; and second, because of compatible new concepts emerging in the major components of the Literary culture: one component embodied in the Division of Humanities, the other in the School of Business (or better of Management), and represented by the long parallel braces.

It is because of the unification of the sciences (nested braces), that inter-translatability not only of scientific, but also of literary and managerial background theories and languages can be effected. This inter-translatability is represented by the convergence of the arrows originating in the left-hand brace (government, business, modern agriculture) which are deeply influenced by the Literary culture (the Division of Humanities) through whose brace the arrows pass (art, philosophy, religion). In the New University these non-scientific aspects of mankind are in harmony with Unified Science, whose nested braces govern the arrows' directions: like theirs, the dominant value-premise of all three sub-cultures is positive.

These three (rather than just two) categories are implicit in C. P. Snow's following statement:

> I gave the most pointed example of this [traditional] lack of communication in the shape of two groups of people, representing what I have christened 'the two cultures.' One of these contained the scientists, whose weight, achievement and influence did not need stressing. The other contained the literary intellectuals. I did not mean that literary intellectuals act as the main decision-makers of the western world. I mean that literary intellectuals represent, vocalize and to some extent shape and predict the mood of the non-scientific culture: they do not make the decisions, but their words seep into the minds of those who do. p. 59.[10]

These minds comprise our system's controllers, the management people represented by the long brace farthest left in Figure IV-11. In the next sentence C. P. Snow returns to his two categories; but he has clearly shown them to be three, with the decision-making managers of industry, government and academe closer to the Literates than to the Scientists, as indicated in our figure: "Between these two groups," Snow concludes, "the [pre-unified] scientists and the literary intellectuals—there is little communication and, instead of fellow-feeling, something like hostility." p. 59.[10] This is, of course, the case in the disintegrating multiversity, Figure IV-9.

A few pages later, however, Snow goes on to "Observe the development of what, in the terms of our formulae, is becoming [in America] something like a third culture" p. 67.[10] This third culture comprizes the center of Figure IV-11, *PUBLIC PHILOSOPHY*, the actual subject of the present book; far greater than the sum of its parts, shown at the left.

When first accepted for publication, this figure did not include the School of Business or Management. Inclusion of this recent addition to higher education was made possible by the brilliant assembly of old and new constructive trends in managerial thought and practice by Carl H. Madden in his new book, *Clash of Culture—*

The Decade of the Seventies. In this book, Madden has formulated the main problems of our civilization. Figuratively speaking, he has put together the lock for which Unified Science is the constructive key. That is to say, his conceptualization is the factor which permits assembly of the actually Three Cultures into the Public Philosophy.

The Marxists had formulated this problem in an incorrect and thus destructive way long ago. In the Depression I had set myself to reformulate the problem, and have been at it ever since. That's how we came to have the key to Madden's lock, ready at the right time; the time which makes its idea irresistible.

Here is the problem as assembled in *Clash of Culture*.[56,57] I quote a few typical passages in a coherent, cumulative way, and add to each passage Unified Science's answer.

> A power structure unsupported by a choate and accepted system of ideas and values is ultimately self-defeating.

Unified Science permits the assembly of a choate system of ideas and values: the Public Philosophy, shown in Figure IV-11.

> The scientific revolution ought to contain within itself the seeds of a grand and radical sense of purpose.

Unified Science contains this seed: the Periodic coordinate system, whose limits are *Alpha* and *Omega*, and whose value-bias is positive.

> 'A new philosophy,' says S. Kristol, 'is needed for capitalism to regain its legitimacy. And that philosophy should be based on the world-view of consistency with the humanist and metaphysical truth of the scientific revolution itself.'

That is the nature of the new Public Philosophy. The result, however, is transformation of both capitalism and socialism into a new system, Social Capitalism, coherently developed in Switzerland from 1925 to the present.[23] (See pages 192–195.) This is the system called for by the Public Philosophy of Higher Industrialism. The old capitalism's legitimacy was lost because—this has been mapped geometrically—its dominant value premise changed from positive to neutral, and later to negative, as described in Ch. V. Social Capital's dominant value-premise however is clearly and unequivocally positive. That is what gives it legitimacy. The new philosophy which Kristol and Madden call for is, of course, essential.

It is, however, the Public Philosophy of Higher Industrial Civilization which results from the transformation of capitalism and of socialism into Social Capitalism.

"We need a discipline rather than a vision," Madden continues.

Unified Science is actually both: it began as a vision—Descartes' dreams, Bacon's *Novum Organum*, Leibniz' Universal Characteristic[12]; perhaps even the sudden picture of the Periodic coordinate system as I walked on Riverside Drive in New York City. Then it developed into the discipline expounded in this book.

"America has led the first scientific revolution, but failed to take the lead in the second."

Does not Unified Science, developed here for thirty-odd years, constitute leadership precisely in the second? America has not failed. This scientific revolution—which involves not merely one, but all disciplines, scientific, humanistic and technological—is bigger than any of the historic revolutions described by Thomas Kuhn. Its period of what he calls *invisibility*[59] (Ch. 11) has therefore been long. Kuhn's analysis of this invisibility's nature, worked out in the United States, permits us now to end this particular invisibility faster than would have been possible otherwise. As this revolution's visibility increases so will America's leadership in this second major scientific revolution.

"Why don't we see the need for simultaneously moving in all fronts which lead to this new [organized] knowledge?"

The creators of Unified Science saw this need decades ago. We are now making what appear as simultaneous moves because three decades of work are presented simultaneously. These now simultaneous moves, called *Unisci*, permit practical solution of our problem.

We have, it seems, designed and produced an early model of the key which Madden calls *policy science*. The next step is to develop it further, and to mass produce it. And Madden points out what is needed: "Corporation resources may be needed to develop policy science."

If these resources are granted, not only the solutions above, but those to scores of other problems called for in *Clash of Culture* will emerge cumulatively, and will transform this culture clash into constructive orientation: in terms of Figure IV-11 it will change the multiversity's *ad hoc* improvisation into rational assembly of "Compatible background theories, coherently mapped into the Periodic coordinate system," resulting in "A single, dynamic, holistic, evolving discipline: the Public Philosophy. This is, and will become more clearly, the Public Philosophy of the Space Age."[34]

Figure IV-11 is a detail-less plan at whose center is a merely synoptic indication of the New University's controller, the New University Council. The point however is, as Stafford Beer emphasizes with heavy type, that **PLANNING IS HOMOLOGOUS WITH ORGANIZATION**.[53] It takes a cybernetic system like the New University, conducted by generalists and organized specialists (Figure VI-12 below), to grasp and deal with whole cybernetic systems. To understand cybernetic systems requires systems-theoretically structured minds and institutions of higher learning. (Hence the invisibility of Unified Science to otherwise excellent scientists.)

What Figure IV-11 shows is such an institution's work component (on the left), its controller (in the center), and its output (on the right). Now let us diagram its output in detail: generalists and organized specialists. (The organized specialist differs from the generalist only in this: one of his input channels is strongly emphasized and articulated.) Since the result of the Council's planning, generalists and organized specialists, is isomorphic (Stafford Beer's "homologous") with the Council's structure, it follows that the Council must consist of organized specialists and generalists.[60]

The generalist's input comes from his field too; but that consists of all the sciences—physical, biological, and psycho-socio-political, Systems-hierarchically organized; and all the humanities—the arts, philosophy, and religion—similarly organized; together with the tactics and strategy of decision-making and execution, of practical leadership, currently called *management*. The latter has been treated in *Plain Truth—And Redirection of the Cold War*.[23] The humanities will be dealt with to a considerable extent in Chapter V. Let us therefore confine ourselves here to the generalist's cognitive aspect, Unified Science.

The generalist's cognitive controller, which Thomas Kuhn calls his *system of paradigms*, organizes his many diverse-appearing kinds of data by means of a single background theory. It maps them coherently into his central paradigm, the Periodic coordinate system. The result is Unified Science.—His output goes to all the sciences and is, *ex hypothesi*, acceptable to them all. We have tested this model for several years in two colleges, two pilot plants, and found it to work as intended. (See Jere Clark's Chapter III.)

The generalist's detector rejects those theories which are incompatible with, and cannot be *made* compatible with, its paradigms.

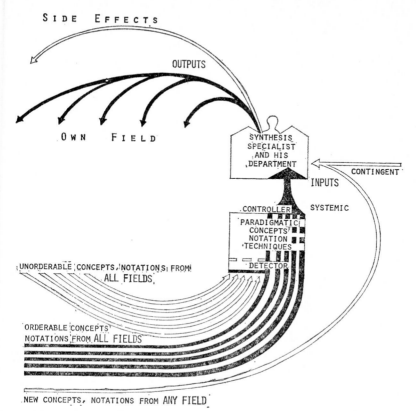

FIGURE VI-12 The Generalist and organized specialist: assemblable products of the New University.

For instance, it rejects the traditional political spectrum; but when this spectrum is corrected and completed, as shown in Chapter II, it is incorporated into the generalist's body-intellectual and body-spiritual.

Such, very briefly, is the New University and its product.

Paternalism still won't work. But now leadership and its complement, followership, will. For the generalist is structured for leadership of the University Council, and the organized specialists for active membership in it. They share the meta-language of Unified Science, and its background theory.

This is what George C. Lodge calls the new ideology.[61] For it includes ecology, and opens the way to its formulation as part of the universal order. And ecology, as Lodge affirms, "has dramatized as no body of knowledge ever has before [but Unified Science has much better since] that everything is related to everything else. It deals a body blow to the individualistic, atomistic view of man espoused by Hobbes and Locke and compels us to concentrate on man as a part of an organic social system, a community, a circle of interrelated facts and elements which are physical and psychological, rational and irrational, technological and spiritual, a circle which in truth is global in scope."[52]

Herewith, our Lower Industrial civilization has come full circle to where it transmutes up to the Higher Industrial Period, or dies. The empirical, inductive, pragmatic mode of thought which shattered the Medieval Circle of Perfection gives rise in the New University to a single, coherent background theory: Industrial civilization's Circle of Perfectibility, the Generalists' and Organized Specialists' consensus. Leibniz' prediction that Unified Science would include ethics, politics, and jurisprudence flows inevitably from his prophetic understanding of the world; his vision of it as what now is called a cybernetic system.[58] For the highest degree of automatization is the highest form of organization: *the control of power by values.* Lodge calls this "the philosophical transformation about which we have no choice—it is happening and there is no going backward."

ARTHUR R. JENSEN
PART II. DIRECT PSYCHOLOGICAL AND GENETIC EMPIRICAL BASIS OF THE PERIODIC TABLE

Scientific progress is won through an unrelenting battle against ad hoc explanations of natural phenomena. The present attempt to bring the behavioral sciences within the purview of Unified Science is thus a welcome and significant step toward understanding behavioral phenomena in terms of a broad conceptual framework that encompasses and unifies also the physical and biological sciences. The periodic and hierarchic schema which forms the essential structure of Unified Science already has a generally acknowledged empirical basis in the physical and biological sciences. Now we must ask, how appropriate is a hierarchical schema in the behavioral sciences? Is there empirical support for thinking of behavior—individual behavior and group behavior—in terms of the proposed system? Probably the best answer that we can presently give to this question must rest upon an examination of the current status of our empirical knowledge in several relevant lines of behavioral investigation. These lines of evidence can best be summarized by posing a number of questions, the answers to which are fundamental to any hierarchical conceptions of adaptive behavior, including Haskell's present formulation.

Phylogeny of Adaptive Behavior.

Are there qualitative as well as quantitative differences in the behavioral adaptive capabilities of animals at different levels of the phyletic evolutionary sequence? That is to say, are there differences not only in the *speed* of learning but also in the complexity of what the organism can learn at all, given any amount of time and training? Are there discontinuities as well as continuities in capacities to perceive, to learn, and to manipulate the environment as we ascend the phyletic scale?

The answer to these questions is now empirically quite clear. There are indeed discontinuities and qualitative differences in learning (i.e. behaviorally adaptive) capabilities as we go from one phyletic level to another. Behaviorally, the phylogenetic hierarchy is best characterized in terms of an increasing complexity of adaptive capabilities and an increasing breadth of transfer and generalization of learning, as we move from lower to higher phyla. It is a fact that every animal, at least above the level of worms, has the capacity to learn; that is, to form stimulus-response associations or conditioned responses. But the degree of complexity and abstractness of what can be learned shows distinct "quantum jumps" going from lower to higher phyla. Simpler capacities, and their neural substrate, persist as we move from lower to higher levels, but new adaptive capacities emerge in hierarchical layers as we ascend the phyletic scale. Each phyletic level possesses all the learning capacities (although not necessarily the same sensory and motor capacities) of the levels below itself in addition to new emergent abilities, which can be broadly conceived as an increase in the complexity of information processing. For example, studies by Bitterman (1965) of animals at various levels of the phyletic scale (earthworms, crabs, fishes, turtles, pigeons, rats and monkeys) have clearly demonstrated discontinuities in learning ability among different species and the emergence of more complex abilities corresponding to the phylogenetic hierarchy. In the experimental procedure known as habit reversal, a form of learning-to-learn in which the animal is trained to make a discriminative response to a pair of stimuli and then has to learn the reverse discrimination and the two are alternated repeatedly, a fish does not show any sign of learning-to-learn (i.e. each reversal is like a completely new problem and takes as long to learn as the previous problems), while a rat improves markedly in its speed of learning from one reversal to the next. When portions of the rat's cerebral cortex are removed, thereby reducing the most prominent evolutionary feature of the mammalian brain, the learning ability of the decorticate rat is exactly like that of the turtle, an animal with little cortex, and would probably be like that of the fish, if all of the rat's cortex could be removed. Harlow and Harlow (1962) have noted similar discontinuities at high levels of learning among rhesus monkeys, chimpanzees, and humans. Again, situations that involve some form of learning-to-learn are most sensitive to differences in capacity. No animals below primates have ever learned the so-called oddity—non-oddity problem no matter how much training they are given, and more complex variations of this type of problem similarly differentiate between rhesus monkeys and chimpanzees. The species differences are not just in *speed* of learning, but in whether the problem can be learned at all, given any amount of training. This is essentially what is meant by a hierarchical conception of learning ability. There is much evidence for this conception, which Jensen (in press) has summarized more extensively elsewhere. The evolution of humans from more primitive forms is now believed to be intimately related to the use of tools and weapons (Ardry, 1961). The mental capabilities involved in the use of implements for gaining ever greater control of the environment, in lieu of sheer physical strength, were just as subject to the evolutionary effects of natural selection as are any genetically mutated organs. More specifically, according to Haskell (1968, p. 475), "What primarily evolves in man is the nerve structure which confers the capacity to invent, to borrow, and to adapt culture traits."

Ontogeny of Human Mental Abilities.

In humans does mental development of the individual occur in qualitatively different stages that are hierarchically related? Are there ontogenetic discontinuities in mental development just as there are phylogenetic discontinuities?

There is now much evidence, originating in the work of Piaget (1960) and substantiated in numerous experiments by other child psychologists both here and abroad (for reviews see Flavell, 1963; Kohlberg, 1968; and Phillips, 1969), that individual cognitive development proceeds by distinct, qualitatively different stages in children's modes of thinking and problem solving at different ages. Piaget and others have demonstrated that children's thinking is not just a watered-down or inferior approximation to adult thinking; it is radically and qualitatively different. The stages of mental development form an invariant sequence or succession of individual development. Each stage of cognitive development is a structured whole; mental development does not consist of the mere accretion of specific stimulus-response associations. Cognitive stages are hierarchically integrated; higher stages reintegrate the cognitive structures found at lower stages. Also, as Kohlberg (1968, p. 1021) points out "... there is a hierarchical preference within the individual ... to prefer a solution of a problem at the highest level available to him." In reviewing the experimental literature on children's learning, Sheldon White (1965) has amassed evidence for two broad stages of mental development, which he labels *associative* and *cognitive*. The transition from one to the other occurs for the vast majority of children between five and seven years of age. In the simplest terms, these stages correspond to *concrete-associative* thinking and *abstract-conceptual* thinking. The latter does not displace the former in the course of the child's mental development; in older children and adults the two modes co-exist as hierarchical layers.

Individual Differences in Mental Development.

Are individual differences in the rate and the asymptotic level of mental development genetically conditioned?

Mental development, as indexed by a wide variety of tests, is known to take place at different rates among children, and the final level of ability attained can be viewed as a hierarchical composite of earlier developed abilities, each level of the hierarchy being necessary but not sufficient for development of the next higher level. At maturity individuals differ with respect to the relative prepotence of different modes in the hierarchy of abilities and thus show differential capabilities for different kinds of learning and problem solving. The difficulty level of items in most standard intelligence tests (especially tests of the culture-fair variety, such as Raven's Progressive Matrices and Cattell's Culture-Fair Tests of g) reflects increasing dependence of the problem's solution upon higher level mental processes.

Over the past half century numerous studies (for reviews see Jensen 1967, 1969) based on a wide variety of tests of mental ability administered to persons of varying degrees of genetic and environmental relatedness, sampled from European and North American Caucasian populations, lead to the now generally accepted conclusion that in these populations genetic factors are approximately twice as important as environmental factors in accounting for individual differences in mental ability. This means, among other things, that variation in mental abilities can be, have been, and are subject to selective pressures of the environment and culture and are subject also to change through different systems of selective and assortative mating, just as is true of physical characteristics that display genetic variation.

Subpopulation Differences in Mental Development.

Are there genetically conditioned differences among population *groups* both in the overall average level of mental development and in the pattern of relative strengths of various mental abilities?

Subgroups of the population which are relatively isolated geographically, culturally, or socially can be regarded as breeding populations to varying degrees (i.e. breeding within groups has a higher occurrence than breeding between groups). To the extent that breeding populations have been subjected to differential selective pressures from the environment, both physically and culturally, differences in gene frequencies can be expected to exist, especially for adaptive characteristics, physical and behavioral, but also for possibly nonadaptive pleiotropic characters (i.e. seemingly unrelated phenotypic effects caused by the same gene). Racial groups and, to a lesser degree, social classes within a society can be regarded as breeding populations.

Social classes as defined largely in terms of educational and occupational status are subject to differential selection for mental abilities. Since these have genetic as well as environmental components, they are transmitted to the offspring, and because of a high degree of assortative mating for mental traits in Western cultures, the gene pools for different social classes will differ in the genetic factors related to ability. The evidence for phenotypic mental ability differences among social classes, along with evidence for genotypic differences, has been reviewed extensively elsewhere (Eckland, 1967; Jensen, 1970). It is now generally accepted by geneticists, psychologists, and sociologists who have reviewed the evidence that social class differences in mental abilities have a substantial genetic component. This genetic component should be expected to *increase* in an open society that permits and encourages social mobility. Phenotypically, of course, social class differences in patterns of mental ability are firmly established. Jensen (1968) has found that lower-class and middle-class population samples differ much less in abilities that are lower in the ontogenetic hierarchy, such as associative learning and memory span, than in higher cognitive abilities, such as conceptual learning and abstract reasoning. A different pattern of correlations between lower and higher abilities also is found in lower-class and middle-class groups, implying a hierarchical relationship among abilities, such that lower-level abilities are necessary-but-not-sufficient for the development or utilization of higher-level abilities.[1]

Scientific knowledge concerning the genetic aspect of ability differences among racial groups, having been generally shunned as a subject of scientific study in modern genetics and psychology, is far more ambiguous and more in dispute than social class differences. The uncertainty in this area will be reduced only through further appropriate research using the most advanced techniques of behavior-genetic analysis. Phenotypically, racial differences in abilities are well established, both with respect to overall average level of performance and to the pattern of relative strengths of various abilities (e.g. Lesser, Fifer, and Clark, 1965). Both social class and racial (Caucasian, Negro, and Oriental) differences have been found in rates of cognitive development as assessed by Piagetian test procedures, such as ability to grasp concepts of conservation of number, quantity, and volume (Tuddenham, 1968). Some indication of the role of genetic factors in the Piagetian indices of levels of cognitive development is shown in a study of Australian aboriginal children, the majority of whom, if full-blooded aborigines, do not show ability for grasping the concepts of conservation of quantity, weight, volume, number, and area, even by the time they have reached adolescence, while the majority of Caucasian children attain this level of mental development by seven years of age. However, aboriginal children having (on the average genetically) one Caucasian grandparent, but reared in the same circumstances as the full-blooded aborigines, performed significantly better (i.e. showed higher levels of cognitive development) than the full-blooded aborigines (De Lemos, 1966).

Personality Correlates of Ability.

Do human behavioral traits other than ability have a genetic component, thereby also being subject to selection, and do such traits become associated, through genetic selection, with intellectual abilities?

Here the evidence is somewhat less well-established than that which was adduced in answer to the previous questions. Eysenck (1967) has amassed extensive evidence for the existence of two broad dimensions or factors of personality, called extraversion-introversion (E-I) and neuroticism (N). The former (E-I) is related to outgoingness and carefreeness; the latter (N) is related to emotional and autonomic instability. Both dimensions have been shown to have physiological correlates and a substantial genetic component, comparable to that found in mental abilities (Eysenck, 1967). Together, these factors, E-I and N, account for most of the individual differences variance in a wide variety of personality assessments. Certain combinations of these traits appear to have socially important consequences. For example, high extraversion combined with high neuroticism is significantly associated with antisocial behavior (Eysenck, 1964).

In a social system such as ours, that tends to sort out people according to their abilities, it seems most likely that those traits of personality and temperament which complement and reinforce the development of intellectual skills requiring persistent application, practice, freedom from emotional distraction, and resistance to mental fatigue and to boredom in the absence of physical activity, should become genetically assorted and segregated, and thereby be correlated, with those mental abilities requiring the most education for their full development—those abilities most highly valued in a technological culture. Thus ability and personality traits will tend to work together in determining individuals' overall capability in the society. R. B. Cattell (1950, p. 98–99) has, in fact, shown that certain personality variables are correlated to the extent of about

0.3 to 0.5 with a general ability factor. Cattell concludes: "... there is a moderate tendency ... for the person gifted with higher general ability, to acquire a more integrated character, somewhat more emotional stability, and a more conscientious outlook. He tends to become 'morally intelligent' as well as 'abstractly intelligent' "

REFERENCES (Chapter IV, Part II)

Ardrey, R., *African Genesis*, New York: Delta, 1961.
Bitterman, M. E., "The evolution of intelligence," *Scientific American*, 1965, **212**, 92–100.
Cattell, R. B., *Personality*, New York: McGraw-Hill, 1950.
DeLemos, M. Murray, *The development of the concept of conservation in Australian aboriginal children.* Unpublished Ph.D. dissertation. University of Western Australia, November, 1966.
Eckland, B. K., "Genetics and sociology: A reconsideration," *American Soc. Rev.*, 1967, 32, 173–194.
Eysenck, H. J., *Crime and Personality*, Boston: Houghton-Mifflin, 1964.
Eysenck, H. J., *The Biological Basis of Personality*, Springfield, Ill.: Charles C. Thomas, 1967.
Flavell, J., *The Developmental Psychology of Jean Piaget*, New York: Van Nostrand, 1963.
Harlow, H. F., E. Harlow, Margaret K., "The mind of man," in *Yearbook of Science and Technology*, New York: McGraw-Hill, 1962.
Haskell, E. F., *Assembly of the Sciences*, **1**, *Scientia Generalis*, 1968 (Xerox).
Jensen, A. R., "Estimation of the limits of heritability of traits by comparison of monozygotic and dizygotic twins." *Proc. Nat. Acad. Sci.*, 1967, **58**, 149–157.
Jensen, A. R., "Patterns of mental ability and socioeconomic status," *Proc. Nat. Acad. Sci.*, 1968, **60**, 1330–1337. (b)
Jensen, A. R., "How much can we boost IQ and scholastic achievement?" *Harvard Educ. Rev.*, 1969, **39**, 1–123.
Jensen, A. R., "A theory of primary and secondary mental retardation." In Ellis, N. R. (Ed.), *International Review of Research in Mental Retardation*, **IV**, New York: Academic Press, 1970.
Jensen, A. R., "Hierarchical theories of mental ability." In B. Dockrell (Ed.), *Theories of Intelligence*, London: Methuen, in press.
Kohlberg, L., "Early education: A Cognitive-developmental view." *Child Development*, 1968, **39**, 1013–1062.
Lesser, G. S., Fifer, G. and Clark, D. H., "Mental abilities of children from different social-class and cultural groups." *Monogr. Soc. for Res. in Child Development*, 1965, **30** (4).
Phillips, J. L., Jr., *The Origins of Intellect: Piaget's Theory*, San Francisco: W. H. Freeman, 1969.
Piaget, J., "The general problem of the psychobiological development of the child." In J. M. Tanner and B. Inhelder (Eds.), *Discussion on Child Development*, **4**, New York: International Universities Press, 1960.

Tuddenham, R. D., *"Psychometricizing Piaget's Methode Clinique.* Paper read at Amer. Educ Res. Assoc., Chicago, February, 1968.
White, S. H., "Evidence for a hierarchical arrangement of learning processes." In L. P. Lipsitt and C. C. Spiker (Eds.), *Advances in Child Development and Behavior,* **2**, New York: Academic Press, 1965.

NOTES AND REFERENCES

1. Haskell, Edward F., with Preface and a chapter by Harold G. Cassidy. *Unified Science—Assembly of the Sciences into a Single Discipline.* Xeroxed, IBM Systems Research Inst., 1969.
2. Posin, Daniel Q. *Mendeleyev, The Story of a Great Scientist,* McGraw-Hill, New York, 1948.
3. Hobhouse, L. C.; G. C. Wheeler; M. Ginsberg, *The Material Cultures and Social Institutions of the Simpler Peoples; an Essay in Correlation.* Chapman, London, 1915.
4. Murdock, G. P., "Ethnographic Atlas," *Ethnology,* Jan. 1962–.
5. O'Connor, Johnson, *English Vocabulary Builder,* Thomas Todd, Boston, 1961.
6. Cambell, J. *The Masks of God,* Viking/Compass, N. Y. 1968.
7. Toynbee, Arnold J., *A Study of History* (Somervell Abridgement of Vols. I–VI) Oxford University Press, N.Y. 1947.
8. To my knowledge no measurement of this most important criterion of intelligence, *abstraction ceilings,* has yet been standardized and tested. The reason probably is that intelligence testers become increasingly vulnerable as their data become increasingly incomprehensible to their clients. Note, for example, the violence directed against conscientious testers who publish careful findings which can offend but cannot be demonstrated to sections of the Majority; violence directed by political activists against scientists such as Arthur Jensen and Richard Herrnstein. This crucially important subject will be dealt with in the concluding chapter.
9. Warner, W. Lloyd and Paul Lunt, *The Social Life of a Modern Community,* Yale Univ. Press, New Haven, 1941.
10. Snow, C. P. *The Two Cultures: and a Second Look,* New American Library, New York, 1963.
11. Period 1 has one class, not *none.* Its controller and work-component are Sub-strata: its controller is the age-group of old people; its work-component is the rest of the age groups.
12. Wiener, P. P., editor, *Leibniz—Selections,* Scribners, New York, 1951.
13. This section has been added since the 1969 symposium, with Dr. Jensen's approval.
14. Pake, George E. "Whither United States Universities?" *Science,* May 28 1971, 908–916.
15. Baltzell, E. Digby, *The Protestant Establishment—Aristocracy and Caste in America,* Random House, New York, 1964.
16. That, history shows, is what one muddles to.
17. Early in 1954 when I was in London, I asked Dr. Toynbee telephonically whether he would approve of an attempt on my part to geometrize his

basic socio-political categories. He cordially approved and wished me success. I began with Geometric Coding of Political Philosophies, proceeded, in collaboration with H. G. Cassidy, with *Plain Truth—And Redirection of the Cold War*, (See below, Note 23) and continue in the present work, with others under way. (See: Haskell, Edward F. "Geometric Coding of Political Philosophies," Proceedings of the Second International Congress of the Philosophy of Science, Vol. IV. Philosophy and Science—History of Philosophy. Editions du Griffon, Neuchatel, Switz., 1955.)

18. I am supported in this belief by Dr. Baltzell's strong endorsement of my position during a heated discussion following a lecture I gave in Philadelphia in 1968.
19. Lippmann, Walter, *The Public Philosophy—On the Decline and Revival of Western Democracy*, Little, Brown, Boston, 1955.
20. Delmos Jones and Steven Polgar conduct a regular feature in *Human Organization*, the journal of the Society for Applied Anthropology; a feature called *Commentary*, which could appropriately have been entitled *Schizm of the Soul*. It provides "a forum for discussion of issues of immediate concern to applied social scientists". Such as, above all, ethical conflicts due to continued failure to codify and internalize psycho-social structure.
   (See, Delmos Jones, Steven Polgar, eds., "Commentary" in *Human Organization*, Journal of the Society for Applied Anthropology. (1703 New Hampshire Ave., N.W. Washington, D.C. 20009) Spring, 1971, 95–101.
21. Williams, George F. *What's It All About?—A Natural Philosophy for our Times*. Exposition Press, New York, 1969, p. 17.
22. Warner, W. Lloyd and J. O. Low, *The Social System of the Modern Factory: The Strike*, Yankee City Series, Vol. IV, Yale Univ. Press, Oxford Univ. Press, 1947.
23. Haskell, Edward F. and Harold G. Cassidy, *Plain Truth—and Redirection of the Cold War*. Offset printed, 1961. Haskell, Edward, "Switzerland's Vertical Front," *Gottlieb Duttweiler*, Speer, Zurich, 1948.
24. Tocqueville, Alexis de, *Democracy in America*, Knopf, New York, 1945.
25. Lloyd Warner was a member of the University of Chicago's Interdivisional Committee to Supervise the work of Edward Haskell (1940–1943). I had studied under him at Harvard and followed him when, in 1937, he transferred to Chicago. I took two or three courses on Yankee City under him.
26. Warner, W. Lloyd and Leo Srole. *The Social Systems of American Ethnic Groups*, Yale Univ. Press, Oxford Univ. Press, 1949.
27. The writer was raised in a missionary family which had, for two generations, oscillated back and forth between the U.S.A. and the Black Sea. They had become citizens of the Atlantic Community long before anyone recognized its existence. The diverse value-systems of the Minority, Inner Majority and Outer Majority; of the various social Strata within these; of the ethnic nationalities comprising these cybernetic system-components; and of the social scientists who study them are set forth in Edward Haskell's *Lance—A Novel about Multi-Cultural Men* (John Day, New York, 1941).
28. Padover, Saul K., ed., *Thomas Jefferson on Democracy*, New American Library, New York, 1946, p. 82.
29. These tests will of course also show the existence and the size of what Thomas Jefferson called the *artificial* aristocracy. "There is also an artificial

aristocracy," he wrote in the same letter, "founded on wealth and birth, without either virtue or talents."[28] We probably can, however, go a great deal deeper: with computer help we probably can show various *degrees* of talents, and *degrees* of diverse temperaments. Also significant *combinations* of these degrees and kinds of innate talent and virtue or viciousness, as the case may be. These we can then match with appropriate schools and other kinds of training.

30. The giant corporations and monopolistic trade unions are fully as disintegrative, intensifying the system's malfunction (see Mintz, Morton, and J. S. Cohen, with a preface by Ralph Nader, "*America, Inc.—*Who Owns and Operates the United States," Dial Press, New York, 1971.)
31. Nicolson, Marjorie Hope, *The Breaking of the Circle—Studies in the Effect of the 'New Science' upon Seventeenth Century Poetry.* pp. XXI, Northwestern Univ. Press, Evanston, Illinois, 1950.
32. "In this Satanic world, the false always appears first and imitates the truth, thus confusing people." Young Oon Kim. (See Kim, Young Oon, *Divine Principle and its Application,* H.S.A.U.W.C., 1611 Upshur St., N.W. Washington, D.C. 20011, 1969.
33. Gierke, Otto von, *Political Theories of the Middle Ages* (translated by E. F. Maitland), Cambridge Univ. Press, London, 1927.
34. Haskell, Edward, "Unified Science: The Public Philosophy of the Space Age," *Connecticut Review,* Board of Trustees for the Connecticut State Colleges, Hartford, Conn., 1969.
35. People displaying this mentality are called Generalists. See *The Moral Society—A Rational Alternative to Death* by John David Garcia. (Julian Press, New York, 1971.)
36. It is true that in feudal societies, royal households, and in Soviet schools for "philosopher kings" not even the first few grades are shared by all Strata. Whether this does or does not further equality of opportunity is a technical question whose answer may require more knowledge than our society yet has.
37. Wiener, Norbert, *Ex-prodigy: My Childhood and Youth,* Simon and Schuster, New York, 1953.
38. Borchard, Ruth, *John Stewart Mill, the Man,* Watts, London, 1957.
39. Kenyatta, Yomo, *Facing Mount Kenya: the Tribal Life of the Gikuyu,* Secker and Warburg, London, 1959.
40. Spuhler, J. N., editor, *Genetic Diversity and Human Behavior,* Viking Fund Publications in Anthropology, New York, 1967.
41. Lippmann, Walter, *The Public Philosophy—On the Decline and Revival of Western Democracy,* Little, Brown, Boston, 1955.
42. Dowling, Colette, "Dropping Out in Manhattan," *New York* (magazine), May 17, 1971.
43. The present unification of the sciences was carried out during a twenty-odd year withdrawal (1948–1969). The symposium whose expanded proceedings are here published initiated the implementation of unified science in an allout attempt to halt the headlong rout of Lower Industrial civilization described by Lippmann, and to transform it into our culture's advance into Period 7, Figure IV-4.
44. Some, of course, do so consciously. E.g. the Soviet leaders who use abundant

tests in screening students for their Minority schools, yet back Americans and Englishmen who denounce the use of tests in similar schools beyond their borders.

45. Kendall, Henry W. and Wolfgang K. H. Panofsky, "The Structure of the Proton and the Neutron," *Scientific American*, June, 1971.
46. Petrie, M. Ann, "Up the Down Campus—Notes from a teacher, on Open Admissions," *New York* (magazine), May 17, 1971.
47. Intensified by Max Ways and his friends under the slogan "More Power to Everybody." (See Ways, Max, "More Power to Everybody," *Fortune*, May, 1970.)
48. Subjective coactions, traditionally distinguished from objective coactions by semi-quotes, are here omitted. Their mapping would not only complicate this figure, but subsume personality concepts which, while touched upon in Chapter II's mapping of political biases, have not been sufficiently elaborated. That occurs elsewhere.
49. Conant, James B. "The Role of Science In Our Unique Society," *Science*, Jan. 23, 1948, p. 78.
50. Sonnemann, Ulrich "The Specialist as a Psychological Problem," *Social Research—An International Quarterly of Political and Social Science*, March, 1951, pp. 9–31.
51. Dr. Ulrich Sonnemann has been a member of C.U.R.E., Inc., since the early 1950s. He is a practicing psychotherapist and resides in Munich, West Germany.
52. Lodge, George C. "Change in the Corporations. Needed: A New Consensus." *Harvard Today*, March, 1972, pp. 6-10.
53. Beer, Stafford, "The Liberty Machine," *Futures*, Dec. 1971.
54. Stafford Beer has been Head of the Department of Operations Research and Cybernetics, The United Steel Companies Limited. He is now President of the Operational Research Society, and Visiting Professor of Cybernetics at Manchester University. This and following quotations are taken from his keynote address to the Conference on the Environment, American Society for Cybernetics, Washington, D.C., October, 1970.
55. Some people, who have not yet grasped the cumulative nature of the System-hierarchy, call it *Post*-Industrial, or *Post*-Technological culture. The term Post, however, implies that in it, industry or technology have ended. This is not, and cannot possibly be the case. Industry must exist, but be controlled, and thus modified, by the new Public Philosophy. Hence the term *Higher* Industrial Culture.
56. Madden, Carl H., *Clash of Culture—The Decade of the Seventies*, (preliminary draft) 1972. Quoted by permission of the author.
57. The following quotations are from a preliminary draft, kindly loaned me by William L. Wallace. The author, who has given his permission, is Senior Economist of the Chamber of Commerce of the United States.
58. Couturat, Louis, *La Logique de Liebniz, d'après des documents inédits*, G. Olms Verlagsbuchhandlung, Hildesheim, W. Germany, 1961, pp. 71,f.
59. Kuhn, Thomas S., *The Structure of Scientific Revolutions*, University of Chicago Press, 1962.
60. This is made possible by an apparently anomalous process popularly called *pump priming*. The New University Council, founded in 1970, consists

of a few generalists and organized specialists who had emerged spontaneously half outside the multiversity. Its core is a group of professors in diverse disciplines who had become friends as undergraduates at Oberlin College in the late Twenties. The others are friends recruited over the years here and there.—The First International Conference on Unified Science, New York, 1972, is being organized around this Council.

61. George C. Lodge is a Harvard Professor of Business Administration.
62. At the end of the school year, however, about 40% of the Freshman class dropped out of their own accord. Contrary to the Dean's prediction, the institution thus seems to be behaving homeostatically. To fulfil his prediction, the college would have to be changed into a high school by changing the faculty.

PART II

1. When the top Strata are included, everything said here is strongly supported and intensified.—See, for instance, Lewis M. Terman, *Genetic Studies of Genius*, Stanford Univ. Press, 1925–59.

Chapter V

Unified Science's Moral Force[1]

EDWARD HASKELL

What Nietzsche has called 'sovereign becoming' is upon us and theory, far from having where to stand beyond it, is chained to its chariot, in harness before it or dragged in its tracks—which, it is hard to tell in the dust of the race, and sure it is only that not theory is the charioteer (Hans Jonas)[2].

We now do have a theory that stands beyond the theory described by Jonas, theory generated by the one-field disciplines. This theory, which stands for Unified Science, represents Nietzsche's "sovereign becoming" itself, and does so cybernetically (see Figure V-1 overleaf).

In the one-field sciences, new theory is based upon (follows) empirical data, and is thus dragged in the chariot's tracks. But it then gives rise to new observations with new instruments and techniques, and through them to new technologies, thus moving in harness in front of the chariot. Soon, however, these new data give rise to improved or radically new theory, the old again dragging the chariot's tracks. Then, being reorganized, theory rushes once more ahead of practice, "in harness before it", generating further new observations and technologies.[3]

Unified Science's theory obviously stands beyond those of the one-field sciences: their theories are part of the data of Unified Science: they are a component of a natural system, human culture, which is in turn part of the world's System-hierarchy, the subject of Unified Science.

No specialized science's theory is, or can be becoming's charioteer for, as Jonas and Figure V-1 point out, "Theory itself [theory of each special science] has become a function of use as much as use a function of theory. Tasks for theory are set by the practical results of

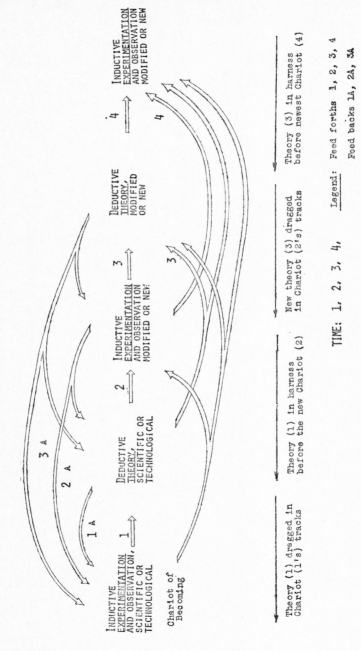

FIGURE V-1 The one-field sciences' reciprocal development of deductive theory and inductive observation and experimentation.—Rightward arrows represent feedforths; leftward arrows, feedbacks.[3]

its preceding use, their solutions to be turned again to use, and so on. Thus [special-science] theory is thoroughly immersed in practice." p. 209.[2]

The aspect of the empirical sciences which is least immersed in practice, the so-called pure sciences, maintain and develop themselves by concentrating on low level systems (such as atoms), or parts of high systems (such as DNA molecules). They develop separate vocabularies, notations, and concept systems, and screen out information about things which they regard as outside their fields, as shown back in Figure IV-10.

This screening results in isolated, discrete bodies of knowledge whose technological applications regularly produce unanticipated and often disastrous "side effects." The input of each science, being discrete from those of other sciences, produces the crisis called the multi-versity, Figure IV-9. Never does any of these sciences clarify or even touch upon the whole truth, most of which is outside any one science's field.

"It is in the realm of concrete judgements and choice that the practical use of theory comes about," Jonas pointed out in 1959. "But this knowledge of use is different not only from the knowledge of the [special science's] theory used in the case, but from that of any [non cosmic] theory whatsoever, and it is acquired or learned in ways different from those of [special science] theory . . . thus there is theory and use of theory, but no theory of the use of theory." p. 199.[2]

This statement, sharpened by inserted qualifications, was true for Hans Jonas in 1959. He then, however, describes Unified Science as it is set forth in Figure V-2: "At the opposite end of the scale is the knowledge concerning ends repeatedly alluded to—of which today [1959] we do not know whether it admits of theory, as once it was held eminently to do. This knowledge alone would permit the valid discrimination of worthy and unworthy, desirable and undesirable uses of science, whereas [fragmented] science itself only permits discrimination of its correct or incorrect, adequate or inadequate, effectual or ineffectual use." pp. 199–200.[2]

Assembling the A, B, C coordinate systems in the central column of the Unified Science Chart (at the rear), and sharpening their coaction cardioids to their logical and geometric points, we obtain the following compass-like representation of Unified Science's frame of reference, the Periodic coordinate system.

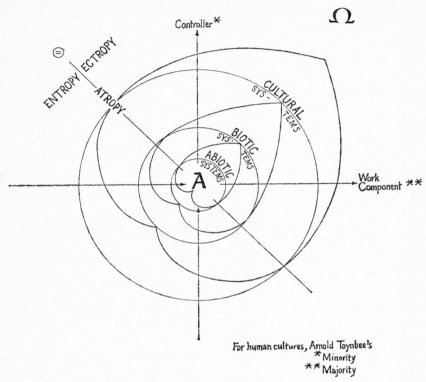

FIGURE V-2. The Coaction Compass defines the relations of the data mapped into it, both to each other and to the Central Order.[3]

The Periodic coordinate system orients any empirical system mapped into it, including the special-sciences' theories, as clearly and precisely as a physical map and compass orient an aeroplane or ship. Unified Science is not immersed in temporal local practices, but immerses them: it shows their directions *sub specie aeternitatis:* it orients their structures by their relations to Å and Ω, the ultimate limits of existence envisaged by modern science and philosophy. Cosmic direction is built into Unified Science as firmly as geographic direction is built into physical compasses. To the extent that the Periodic coordinate system shows what Werner Heisenberg calls "our relationship with the central order", it can in the final analysis be our compass.

Do not Figures V-1 and V-2 show that this knowledge, intuited long ago by Plato and now by Jonas in Literate metaphors, has come full circle, and reappears here in the geometric language and idiom of the Higher Industrial culture?

"Obviously," Jonas points out, "it is a different kind of knowledge that has to do with the desirability of ends, and a different kind that has to do with feasibility, means, and execution." p. 198.[2] Figure V-2 elucidates this difference: the data of the special sciences and technologies (data mapped into the Periodic coordinate system by means of their characteristic numbers, as for instance in Figures II-14a and b) have to do with feasibility, means, and execution. The process of mapping these data into the Periodic coordinate system shows their desirability or undesirability in terms of their vector-direction: if they transmute the system over-all and in the long run (diachronically) toward Å they are evil and undesirable; if toward Ω they are desirable and good. And every human being who loves life and seeks its preservation and advancement will concur. The acquisition of the data, which occurs by means of special sciences, results in knowledge of feasibility, means and execution; the orientation of this knowledge by means of the Periodic coordinate system results in knowledge of the desirability of ends. The two together constitute our map and compass.

1. UNIFIED SCIENCE IS SCIENCE

The condition for Jonas' two different forms of knowledge to coexist and further each other is that they be mutually compatible, and that both be amenable to empirical verification or disproof. Both of these conditions are fulfilled by Unified Science, an assertion which is sure to be put to adequate tests.

It has, in fact, been tested constantly during the thirty-odd years of its development. During this time, Unified Science has itself undergone the cybernetic kind of development diagrammed in Figure V-1. Its nature, however, has determined the *direction* of its development: this direction has consisted in the kinds of change that physical compasses have displayed: the Periodic coordinate system has become ever more precise, versatile, reliable, and widely applicable. For where the nature and purpose of an instrument is normative—as it is in compasses, whether physical or moral—increase of normativity is the built-in direction of the instrument's development. By the same token, where a theory's nature and purpose is the discovery and refinement of relationships *sub specie localis*, as are those of the one-field sciences (Figures IV-9, 10), its built-in mode of development is an increase of analytical detail, and its cosmic direction is unascertainable except relative to a normative frame of reference.

"In its positive aspect", Jonas out, "good will is for the good and must therefore be informed by a conception of what is good . . . If there is a knowledge of it, not [one field] science can supply it. Mere benevolence cannot replace it—nor even love, if love without reverence; and whence can reverence come except from a knowledge of what is to be revered?" Then comes the paralyzing portent of despair: "But even if a guiding knowledge of the good, that is, true philosophy were available, it might well find its counsel to be of no avail against the self-generated dynamics of science in use." p. 197.[2]

From where does Unified Science derive its concept of good and evil? From the customs of a given time and locality? Certainly not. From ancient manuscripts or books? Not at all—though its concepts agree with and confirm certain of these. From mystical visions and enlightenment? Yes, in the sense that flashes of insight, wonderful dreams, visions and peak experiences have been the lot of scientists in every field.[5] Peculiar to Unified Science is the convergence of insights from all fields: all fields of science, first of all; but also from literature, music, and art and also from religion, the synthesis of humankind's non-scientific experience. Speaking in Unified Science's own terms, its conception of good and evil is derived from and confirmed by the structure and behavior of every Group, Stratum, Period, Major Stratum and Major Period in the universe. Its conception of good and evil is informed in thousands of empirical, mutually reinforcing ways. Its knowledge of evil and good is indeed not supplied by any one-field science, but by their assembly into a co-ordinate system whose axes are related to the world's universal directions of change, entropic, atropic, and ectropic; directions derived from the assembly of cosmic data and ascertainment of their change-directions.[6]

A powerful moral force is generated by this organization of knowledge, and by incessant empirical verifications of the moral law they manifest: $R = f(\theta)$.

Jonas naturally assumes that even if an acceptable normative discipline were to emerge it would lie outside the domain of science and therefore could not control its development. This assumption leads him nearly to despair: "The effecting of changes in nature, as a means and as a result of knowing it are inextricably interlocked," he declares. (And they are thus diagrammed in figure V-1.) "And once this combination is at work it no longer matters whether the pragmatic destination of theory is expressly accepted (for example by the 'pure' scientist) or not. The very process of attaining knowledge leads through manipulation of things to be known, and this origin fits of itself the theoretical results for an application whose possibility is *irresistible*—even to the theoretical interest, let alone the practical, whether or not it was contemplated in the first place." p. 205, italics mine.[2]

Jonas, however, is contradicted by the unexpected fact that Unified Science turns out to be *science*. (It is, of course, also philosophy, as shown below.) The fact is that Unified Science itself, as well as its components, are empirically and logically verifiable. What happens appears clearly in terms of Figure V-1 when it is altered by the following substitution: for *Deductive Theory*, we substitute *One-field Disciplines* (Figures IV-9, 10). What happens next is suddenly new and unexpected: the Periodic coordinate system is anchored to ultimate and unchangeable absolutes, Å and Ω, absolutes generated by science itself; absolutes which science can approach but cannot replace or change essentially. Unified Science can be refined and elaborated by the empirical sciences. But because it is science, it can and does do what Jonas believed could not be done: it can coordinate and morally direct the one-field sciences' development (Figures IV-11, 12).

A basically new kind of system has herewith emerged; one whose unalterable norms, discovered by science itself, scientists are bound by their own rules, standards and principles to respect and observe. Unified Science can therefore check and redirect the cancer-like growth of the one-field sciences, technologies and ideologies whose terrible and here-to-fore uncontrollable power Jonas has so brilliantly described.

Some such developments are already being redirected: America's vast Super Sonic Transport project, for instance, was scrapped by the United States Government at enormous financial cost, and a powerful revival attempt was later resisted even by its manufacturers themselves. Dr. George T. Lodge has consolidated this position by demonstrating the preponderance of undesirable over desirable consequences of Super Sonic Transport.[7] What we see here emerging is the *technological ombudsman* which Alvin Toffler described in *Future Shock*.[8] Computer simulation of webs-of-life and webs-of-mind (Chapters II and IV) will probably become its most effective method.

The time will soon come when no proposed technological innovation will be executed until it has been subjected to moral analysis; that is, oriented relative to the Periodic coordinate system and judged to be constructive and ectropic over-all. May not, then, Unified Science be called "sovereign becoming's" charioteer?

2. UNIFIED SCIENCE IS MODERN PHILOSOPHY IN THE ORIGINAL GREEK SENSE.

"To the Greeks," says Jonas, "be it Plato or Aristotle, the number of the truly knowable things is finite, and the apprehension of first principles, whenever obtained, is definitive—subject to intermittent renewal but not to obsolescence through new discovery and better approximation." pp. 206–207.[2]

The present book affirms in its title that with its appearance, mankind has come full circle. It claims that to nearly all scientists—one-field specialists and Unified-Scientists alike—the point of maximum entropy A and the region of maximum ectopy Ω are incapable of obsolescence through new discovery and better approximation; and that the number of truly knowable systems is not infinite but finite. Until these limits were conceived and the Systems-hierarchy extending between them was defined, it had been, as Hans Jonas says, "inconceivable to the modern experience of knowledge . . . that any state of theory, including the conceptual system of first principles governing it, should be more than a temporary construct to be superseded by the next vista to which it opens the way when all its implications are matched againstall the facts." pp. 206–207.[2]

Unified Science's absolutes are permanent. So future shock—the fear that permanence is dead—has lost its sting.[8] Jonas' book, from which I have been quoting, bears a predictive sub-title: *Toward A Philosophy of Biology*. In its Epilogue, he defines the objective toward which his work is directed, and clearly points the road by which it has here been reached: "Ontology as the ground of ethics was the original tenet of philosophy. Their divorce, which is the divorce of the 'objective' and 'subjective' realms, is the modern destiny. Their reunion can be affected, if at all, only from the 'objective' end, that is to say, through a revision of the idea of nature. And it is becoming, rather than abiding, nature which would hold out any such promise."

Unified Science is a profound revision of most peoples' idea of nature; and it is stated in terms precisely of process, of becoming. The insight which Jonas then shows us is prophetic: "From the immanent direction of its total evolution there may be elicited a destination of man by whose terms the person, in the act of fulfilling himself, would at the same time realize a concern of universal substance." Unified Science proclaims this immanent direction to be increasing organization; and its destination, ectropy's highest region, to be Ω. "Hence," Jonas continues, "would result a principle of ethics which is ultimately grounded neither in the autonomy of the self nor in the needs of the community, but in an objective assignment by the nature of things (what theology used to call the *ordo creationis*) such as could still be kept faith with by the last of a dying mankind in his solitude." p. 283.[2] Some call this principle *Omega*, others call it *God*.

In writing his book, Jonas was moving toward the kind of structure which Walter Lippmann has called the *Public Philosophy*,[9] the structure which is shared by all great religions and ideologies. This structure has remained constant through all of the public philosophy's changes, from the animism of Lower Hunters, Period 1; through the Great Religions of the Literates, Period 5; to the emergence of the Higher Industrialists, Period 7, which the unification of sciences into a public philosophy permits and requires. In Figure IV-11 we have diagrammed this permanent ideocratic structure.

Every public philosophy encompasses vocabularies, observations and theories from all its culture's fields; that is to say, all those which its paradigmatic concepts can encompass and order. All others, each public philosophy must and does screen out.[10] Whenever this ceases to happen, the civilization breaks down *ipso facto*: its controller has broken down.

Hence the first part of the sub-title of Lippmann's exceedingly important book: "On the Decline and Revival of Western Democracy." The two parts of Lippmann's sub-title correspond to our civilization's two alternatives. Our Lower Industrial Period (Figure IV-I) is, in any case, a transitional phenomenon; and it can develop in just two directions: either it will continue to *decline*, turn out to be a cultural abortion—a cultural embryo which, never having succeeded in generating its public philosophy has, during this century, been expiring as history's most despicable garbage heap, destined at last to consume itself by internal combustion. Or else *reviving*, it will turn out to have been generating its public philosophy successfully and will, with its help, transmute into predominantly positive coactions and consolidate our planet's climax ecosystem, the Higher Industrial Period.

My prediction has always been, and is, that we will make it to the top! In 1942 I announced *The Religious Force of Unified Science*.[11] Then—after Lippmann had formulated the more general concept, *public philosophy* (1959), and Unified Science had fully emerged—the position became clear and specific: *Unified Science: The Public Philosophy of the Space Age*.[12,13] How do we know that Unified Science, and it alone, can be our public philosophy? This is the Age of Sciences; and more particularly, of systems-theoretic sciences; and Unified Science coordinates sciences in terms of general systems theory.[14] Being systems *science*, it admits empirical data. Being *unified*, it admits them from all disciplines and fields of experience. (Some of its multiple inputs are specified on the left-hand side of Figure IV-11, *Proposed Transformation of Multiversity into University*.) Being steered by a coordinate system whose axes represent the General System's work component and controller (Figures II-10 and chart), it orders all these data relative to all their theoretically possible coactions, and these relative to the world's ultimate absolutes, Å and Ω. Ordering all knowledge and experience relative to Å and Ω, it shows clearly the universe's strongly positive value-bias. Showing a positive value-bias, it corresponds unequivocally to the positive value-premise of all the great religions.[15] And having done this, Unified Science has come Full Circle: it supports, and is supported by, the public philosophies of all the great Literate civilizations, past and present, and the majority of our own civilization's people.[16] And doing this in the language and idiom of science, it bears our civilization's signs of legitimacy and seals of truth.

Is this not reversal of what Toynbee called *schizm of the soul*? Is it not reversal of the disaster which causes, and results from, the *schizm of the body politic, schizm of the body ecologic*, and *disintegration of civilization*? Toynbee calls schizm-reversal *transfiguration* and *palingenesia*.[16] In *The Public Philosophy*,[9] Walter Lippmann describes what, for our civilization, palingenesia would be: "A convincing demonstration . . . that the principles of the good society are not, in Sartre's phrase, invented and chosen—that the conditions which are to be met if there is to be a good society are there, outside our wishes, where they can be discovered by rational inquiry, and developed and adapted and refined by rational discussion." He then concludes as follows:

"If eventually this were demonstrated successfully (as I affirm it can) it would . . . rearm all those who are concerned with the anomy of our society, with its progressive barbarization, and with its descent into violence and tyranny. Amidst the quagmire of moral impressionism they would stand again on hard intellectual ground where there are significant objects that are given and not merely wished. Their hope would be re-established that there is a public world sovereign above the infinite number of contradictory and competing private worlds. Without this certainty, their struggle must be unavailing... For political ideas acquire operative force in human affairs when, as we have seen, they acquire legitimacy, when they have the title of being right which binds men's consciences. When they possess, as the Confucian doctrine has it, 'the mandate of Heaven.'

In the crisis within Western Society, there is at issue now the mandate of Heaven." pp. 180–181.[9]

What is the form which this mandate must today assume? In our Scientific-Industrial civilization, the title of being right which binds men's consciences bears the sign of logical consistency and the seal of empirical validation. Some of the sciences' traditional terms of discourse have now, I believe, been altered in the manner which Lippmann advocates, producing the result he intuited and expressed more movingly than any scientist. However, belonging to the Literate Culture as he does, he could point out but could not himself lead the way into the Promised Land. The recreation of public philosophy in our society must be led by the keepers of its signs and seals of legitimacy—by scientists. They are achieving it by converging the sciences into a single discipline. For this is the Public Philosophy of the Space Age. This they can achieve because, and only because, with all its novelty and power, it is in basic harmony with the old Literate public philosophy which the fragmented sciences had brought into decline. The major forces of disintegration will, under this mandate of Heaven, be turned around and organized constructively.

3. THE BUCK STOPS AT ALPHA

In Chapter I, Harold Cassidy affirms that there has been discovered "a scientifically based pattern of a universal kind which is displayed in some respect by all of human knowledge and experience of nature and man... If I say that, in my opinion this pattern... constitutes an *invariant relation* that enables translation between various developing fields of knowledge and experience, then at least metaphorically one can understand me to mean that, like the Lorentz Transformations, it makes the applicable relativity tolerable."

What is an invariant relation? It is a relation such that, as the French saying has it, *plus ça change, plus c'est la même chose*. The more things change, the more they remain the same. That sameness—as we move up the changing System-hierarchy, Figure II-1—is the universal pattern which Cassidy discerns; the major paradigm of Unified Science, Figures V-2 and V-5.

In order for this simple sameness to be repeated throughout the richly strange universe, there must be in existence a set of relational variables which all systems have in common. Moreover, there must be a totality of sets, and a relation among this totality's parts, such that all *systems* of the Systems-hierarchy display it, *too*, in common. This constant relation among the sets of relational variables is the invariant that all things must have in common for Cassidy's statement to be correct. It must exist, however, on a higher level than they, for it is their higher derivative, the constancy of change which Jonas calls *becoming*.

The next situation which is now upon us is made clear in Willard V. Quine's incisive exposition of what has here-to-fore been ontology's apparent relativity.[46] This relativity is inherent in the one-field sciences' discrete studies of systems parts *sub specie localis*, Figure V-1, and is the transitory basis of existentialism, cultural relativity, promiscuity, and negative and zero value-biases, the modern forms of *schizm of the soul*.

"Ontological relativity," Quine points out, "is not to be clarified by any distinction between kinds of universal predication—unfactual and factual, external and internal. It is not a question of universal predication. When questions regarding the ontology of a theory are meaningless absolutely, and become meaningful relative to a background theory [such as, for instance, that of Unified Science], this is not in general because the background theory has a wider universe . . .

"What makes ontological questions meaningless when taken absolutely is not universality but circularity. A question of the form 'what is an F?' can be answered only by recourse to a further term: 'an F is a G.' The answer makes only relative sense: sense relative to the uncritical acceptance of 'G'." p. 53.[46] What Quine means thus appears to be infinite regress, of which circularity is a special case.

In Unified Science the buck called *infinite regress* stops at Alpha Å. What is a molecule? A combination of atoms. What is an atom? A combination of particles. What is a particle? A combination of partons. What is a parton? We are in process of answering that

Unified Science's Moral Force   181

question now. But we are no more caught in infinite regress than are physicists in attempting to reach absolute zero, 0°K. We may never reach Alpha, as they may never reach 0°K. But what we and they are approaching are not infinities, but absolute limits.

So also in the opposite direction: what is a geoid system, a cell, a plant, an animal, a man? Each of them is a combination of systems anterior to itself in the System-hierarchy.[18] Again we are no more caught in infinite regress than are physicists in attempting to reach the highest temperature or speed. We may never reach Omega Ω. What we, however, are approaching is not infinity, but an absolute and permanent limit.

Unified Science avoids infinite regress by deploying the one-field sciences' data System-hierarchically from Ω, through A to Anti-Ω, (Figures II-10 and the fold-out chart). The former meaninglessness of ontological questions resulting from infinite regress is hereby precluded.

Also precluded are the major ambiguities which Quine groups under the *inscrutability of references* p. 38. One aspect of this inscrutability arises (in English and other traditional languages) through failures to distinguish a system (say a rabbit), its undetached parts, and its stages. Unified Science overcomes this source of confusion by means of its Characteristic Numbers. In the web-of-life, Figure II-14b, the rabbit is defined as $2\genfrac{}{}{0pt}{}{4}{4}6$. This Characteristic Number specifies an organism belonging to the animal kingdom (Major Stratum 6), belonging to the gathering and hunting animal Period (lower 4) and Stratum (upper 4), and currently in the ontogenetic stage of infancy (2 at the left). Undetached rabbit parts are excluded by stating the Major Stratum symbol (6 in the center) which specifies the kingdom of animals.

There remains now what Quine calls *the problem of ostension*. The terms of empirical science depend upon two kinds of pointing or ostension, direct and deferred. Quine defines *direct ostension* thus: "The *ostended point*, as I shall call it, is the point where the line of the pointing finger first meets an opaque surface. What characterizes *direct ostension*, then, is that the term which is being ostensively explained is true of something that contains the ostended point. Even such direct ostension has its uncertainties, of course, and these are familiar. There is the question how wide an environment of the ostended point is meant to be covered by the term that is being ostensively explained. [In Unified Science this is specified by the Period number.] There is the question how considerably our absent thing or substance might be allowed to differ from what is now ostended, and still be covered by the term that is being ostensively explained. Both of these questions can, in principle, be settled as well as need be by induction from multiple ostensions . . . " pp. 39–40.[46] The multiple parts of Characteristic numbers are multiple deferred ostensions.

As for *deferred ostension*, it is ostension by inference: "It occurs when we point to the gauge, and not the gasoline, to show that there is gasoline. Another such example is afforded by the Gödel numbering of expressions." p. 40.[46] For instance, the delayed ostension involved in relating the four parts of Characteristic numbers, as expounded in Figure II-14, to the languages of the one-field sciences, and through these to the objects and processes of nature.

Now comes the crunch, what Quine calls "the perennial philosophical problem of induction". Only solution of this problem can transfigure the ontological relativism characteristic of the precarious Lower Industrial Period's civilization into Lippmann's "convincing demonstration . . . that the principles of the good society . . . are there, outside our wishes, where they can be discovered by rational inquiry and developed and adapted and refined by rational discussion." [9]

"To trust induction as a way of access to the truths of nature", Quine points out, "is to suppose . . . that our quality space matches that of the cosmos. The brute irrationality of our sense of similarity, its irrelevance to anything in logic and mathematics, offers little reason to expect that this sense is somehow in tune with the world— a world which, unlike language, was never made. Why induction should be trusted, apart from special cases such as the ostensive learning of words, is the perennial philosophical problem of induction". pp. 125–6.[46]

This problem's solution is implicit in the structure of the Systems-hierarchy: the sense of similarity is clearly displayed in the formation even of atoms, and is displayed in the formation and behavior of all natural systems classed in each of this hierarchy's Major Periods thereafter, the human mind included. Protons display this sense of similarity in constantly orbiting one natural kind of thing, electrons; and electrons in constantly selecting two kinds of things (Ontological Relativity Chapter 5), nuclei composed of protons and neutrons, around which to orbit. And so forth through all the abiotic, biotic and cultural structures and systems of the universe up to the human mind, logic and mathematics most decidedly included. All these display what strikes us as the *rationality* of our sense of similarity, rationality of which logic and mathematics are the highest known expressions. Does not the universe's teleonomic *Werden* or becoming—culminating in language, mathematics, and logic—give us every reason to expect our sense of similarity to mesh with the senses of similarity of all systems antecedent to ours in the Systems-hierarchy? This is the natural conclusion which springs from what Jonas has called for and Unified Science is based upon: revision of the idea of nature, and of its becoming.

What could it be but our heretofore disparate and unassembled idea of nature—including the traditional separation of the study of logic and mathematics from studies of antecedent parts of the System-hierarchy—that makes the brute (and *a fortiori* the plant, geoid system, molecule, atom and particle) appear irrational; that sees our sense of similarity as "brute irrationality"; and makes it appear irrelevant to its own highest descendants, the hierarchy's highest components, logic and mathematics?

When these traditionally separate studies are unified, as they are now, Man's mind, including its ostensive coining of words, is recognized as the System-hierarchy's highest known Major Stratum, the Kingdom of Man. It is then evident that to the extent that the structure and contents of Man's mind consist of Unified Science, its quality space must, and does, match that of the cosmos; and that it does so both logically and empirically. Through geometric synthesis of the multiple ostensions of the one-field sciences, the coining and learning of words (the special case in which the one-field specialist trusts induction) is generalized into the broadest system of deferred ostensions, Unified Science; "the universe writ small."

We do, of course, "need a background language," as Quine says, "to regress into." And here again the question arises, "Are we involved now in an infinite regress?" He answers it thus: "If questions of reference of the sort we are considering make sense only relative to a background language, then evidently questions of reference for the background language make sense only relative to a further background language. In these terms the situation seems desperate, but in fact it is little different from questions of position and velocity. When we are given position and velocity relative to a given coordinate system, we can always ask in turn about the placing of origin and orientation of axes of that system of coordinates; and there is no end to the succession of further coordinate systems that could be adduced in answering the successive questions thus generated." p. 49.[46]

This is the unsatisfactorily incomplete kind of answer to which one-field specialization in formal science leads. Immediately, therefore, Quine relates it to other parts of the System-hierarchy.

"In practice of course we end the regress of coordinate systems by something like pointing." And that something is mapping: equating origin and axes to empirical phenomena which are, of course, hierarchically organized. "And in practice," Quine goes on, "we end the regress of background languages, in discussions of reference, by acquiescing in our mother tongue and taking its words at face value." Once, that is, the concepts of these words' denotata have been organized System-hierarchically, as they are in Unified Science.

"Very well," he goes on, addressing himself to formal specialists. "In the case of position and velocity, in practice pointing breaks the regress. But what of position and velocity apart from practice? What of the regress then? The answer, of course, is the relational doctrine of space; there is no absolute position or velocity; there are just the relations of coordinate systems to one another, and ultimately of things to one another. And I think that the parallel question regarding denotation calls for a parallel answer, a relational theory of what the objects of theories are. What makes sense is to say not what the objects of a theory are, absolutely speaking, but how one theory of objects is interpretable or reinterpretable in another." pp. 49–50.[46]

The point of the Unified Science enterprize is to organize the one-field disciplines so that the objects and theory of each discipline are interpretable or reinterpretable in terms of those of the others, Figures IV-11, 12.

Quine describes the method here employed as follows: "The reduction of one ontology to another with help of *proxy function:* a function mapping the one universe into part or all of the other." p. 55.[46] This kind of mapping has been the principal activity throughout this book as shown, for example, in Figure II-1 and the Unified Science Chart.

The language and theory developed in this process, what Quine calls the *background language* and *background theory* of the one-field disciplines, are *ipso facto* the language and theory of Unified Science. "Our dependence upon a background theory," he says, "becomes especially evident when we reduce our universe U [the universe of one scientific discipline] to another V by appeal to a proxy function.

Unified Science's Moral Force

For it is only in a theory with an inclusive universe, embracing U and V [plus all the other scientific disciplines], that we can make sense of the proxy function. The function maps U into V and hence needs all the old objects of U as well as their new proxies in V." p. 57.[46]

Partial syntheses—syntheses of groups of sciences such as physics, chemistry and biology, or psychology and sociology—do not resolve the ontological problem. Quine implies this as follows:

"All that is required toward a function is an open sentence with two free variables, provided that it is fulfilled by exactly one value of the first variable for each object of the old universe [one discipline] as value of the second variable [another discipline]." Such a function is implicit throughout the System-hierarchy. It is implied, for instance, in Figure II-1. "But the point is that it is only in the background theory, with its inclusive universe, that we can hope to write such a sentence and have the right values at our disposal for its variables." p. 58.[46] That is the theory of Unified Science.

How can the small human mind presume to grasp the enormous universe as a whole, as Unified Science affirms it can? Quine answers this question by way of the Löwenheim-Skolem theorem as follows: "It says that if a theory is true and has an indenumerable universe, then all but a denumerable part of that universe is dead wood, in the sense that it can be dropped from the range of the variables without falsifying any sentences.

"On the face of it," Quine points out, "this theorem declares a reduction of all acceptable theories to denumerable ontologies. Moreover, a denumerable ontology is reducible in turn to an ontology specifically of natural numbers, simply by taking the enumeration as the proxy function, if the enumeration is explicitly at hand. [As, for instance, in the Periodic table of chemical elements.] And even if not at hand, it exists; thus we can still think of all our objects as natural numbers, and merely reconcile ourselves to not always knowing, numerically, which number an otherwise given object is." p. 59.[46]

Ontological relativity's "infinite" regress, including its circularity, has herewith been shown to be finite, its indenumerability shown to be irrelevant, its multiple ostensions codified, its deferred ostensions referred to the Systems-hierarchy, its "inscrutability" of reference made scrutable, its many background theories reduced to a meta-background theory.

How has this come about?

It came about by reversing some, and changing some of the tacit assumptions or paradigms which underlie the unorganized, and thus malignant growth of one-field specializations comprising the multiversity. It came about during some thirty-odd years of scientific revolution.

4. A COPERNICAN CHANGE OF ATTITUDE

"The key to the progress of the natural sciences in Europe [and thus to the rise of the Lower Industrialist out of the Literate culture], lay very largely in a growing habit of testing theories against careful measurement, observation, and upon occasion, experiment," the historian William McNeill points out. "Astronomers and physicists undertook closer observations and more exact measurements only after Copernicus (d.1543) had put an alternative to traditional Ptolemaic and Aristotelian theories before the learned world; and Copernicus did so, not on the basis of observations and measurements, but on grounds of logical simplicity and aesthetic symmetry." p.593.[17]

Today, Aurelio Peccei, president of the scientific Club of Rome, is calling for "A Copernican change of attitude".[19] For the modern industrial world, this change of attitude is presented in the book you hold before you now: it is an alternative to our malfunctioning congeries of specialized one-field scientific and humanistic theories. We present this alternative to you on the basis of observations and measurements, such as Arthur Jensen's, and on the grounds of logical simplicity. And, far beyond these grounds, we present it because this Copernican change of attitude is necessary to the survival of the Empire of Man and of its otherwise doomed participants—human, animal, vegetable, and mineral.

The universal pattern displayed in Unified Science has existed right along. Why, then, has it not been obvious ever since the modern sciences emerged, four centuries ago? For the same reason that it was not obvious, until Copernicus, that the Earth revolves around its axis and the sun.

"Consider," says Thomas Kuhn, "the men who called Copernicus mad because he proclaimed that the Earth moved. They were not just wrong or quite wrong. Part of what they meant by 'earth' was fixed position. Their earth, at least, could not be moved. Correspondingly, Copernicus' innovation was not simply to move the earth. Rather, it was a whole new way of regarding the problems of physics and astronomy, one that necessarily changed the meaning of both 'earth' and 'motion.' Without those changes the concept of a moving earth was mad." pp. 148–9.[10]

So with the discovery of universally invariant relations which form the basis of science-synthesis. It involves a whole new way of regarding the problems of the physical, biological, and psycho-sociopolitical sciences.[20]

This is the kind of concrete, verifiable change of theory which epitomizes a Copernican change of attitude and permits empirical scientists to line up for and against this scientific revolution.

The scientific community has been groping for decades in the invariable precursor of a scientific revolution: deepening crisis.[10] Today, as in the past, the crisis has called forth a host of immature, competing theories; in our case, attempts at scientific synthesis. However, since this crisis extends far beyond the sciences—since it was brought on by the cancer-like proliferation of separate and independent disciplines, scientific and humanistic (Figure IV-9) and its resolution demands *synthesis*—the synthesized theories cannot be just scientific. They must include the ancient literary tradition in whose terms the spiritual syntheses of pre-industrial Strata and Periods have always been, and must today continue to be couched: the language of arts, religions and philosophies.[21]

Calls for up-dating and synthesis of religions, which is essential to our industrial civilization, have been voiced by powerful theologians: "Christianity without Religion" by Dietrich Bonhoeffer (who was hanged by the Nazis), and "The Humanity of God" by Karl Barth in Germany; "Depersonalized Religion" by the Jesuit scientist Teilhard de Chardin in France; "Honest to God" by the Anglican Bishop John A. T. Robinson in England;[22] "The Shaking of the Foundations" by Paul Tillich in America; Sun Myung Moon's "Divine Principle" in South Korea, Japan, and by now many other nations.[23]

John A. T. Robinson, and others who call for an up-dating of religion, being theologians and thus one-field specialists themselves, could understandably not specify just how to do it. The consequent uproar was therefore unproductive theological debate: "The Honest to God Debate".[24] It had to be a man who combines in his personality the literary with the scientific culture, who foresaw and predicted the nature of the brewing revolution: C. P. Snow. Snow stated in 1963 that the Two Cultures, scientific and literary, were about to come together into a single entity, and even predicted where: in the United States.[25]

Though we had corresponded briefly, and I had publicly discussed "The Religious Force of Unified Science",[11] he did not say that Unified Science would, by its definition and intrinsic structure, have to be this sort of synthesis. Nor did he say that the public philosophy of Higher Industrial Civilization can scarcely be anything but this over-all sort of synthesis. Nonetheless, he sensed and said clearly that the Two Cultures were about to unite. This prediction is now being fulfilled. And as this happens it will restore in our New World the ancient oneness of the human spirit.[26]

In 1955 Walter Lippmann hoped and pleaded for this restoration in *The Public Philosophy—On the Decline and Revival of Western Democracy*. I had replied that Unified Science is The Public Philosophy of the Space Age.[12] And this involves a global, and thus Copernican, change of attitude.

This general synthesis presupposes, and depends upon, paradigm-changes such that the paradigms of all the relevant views and disciplines become compatible. And that is clearly what is happening today: each of the paradigms of Unified Science involves a profound change of its predecessor. And all of them combine into a coherent and largely verifiable system; a system of attitudes which tends to support the paradigms of advancing theologians, political scientists, economists, and men and women of the arts and literature.

The paradigms of Unified Science read as follows:

I. The universe is building up as well as running down.[48]

II. Phenomena which seem empirically different are basically similar.

The empirical differences which led early scientists to divide their studies of the universe's Major Strata into separate, independently developing disciplines—and beyond these, into the West's Two Cultures—are recognized in Unified Science as less important than the invariant relation which it shows them to have in common.

III. Natural systems [the simplest ones, of course, excepted] are fundamentally inclusive.

The mutual exclusiveness of traditional scientific categories—such as the kingdoms of man, of animals, of plants, or of atoms—is of course recognized and honored. (Thought itself requires that it be.) But Unified Science affirms the strategic prepotency of inclusiveness, demanded by understanding of the System-hierarchy. The basic unit of Unisci theory is the *system*, not one or another of its parts. (Those are the main thought units of unorganized one-field specialists.) Unified Science affirms as axiomatic that every natural system except the lowest *includes* systems lower than itself in the System-hierarchy. Inclusiveness is implicit in Unisci's fifth paradigm: this axiom affirms the preponderance of positive coaction. And its acceptance as a strategic attitude involves our thought-parts in positive, moral coaction over-all.

IV. The lowest members of the System-hierarchy are certainly teleomorphic and teleonomic, and may be teleological.

The hierarchy's latest and highest members' structures and behaviors, which appear to be its end or goal (*telos*) are implicit in, and generally determined by, the hierarchy's earlier and lower members.

Since this hierarchy's highest known emergent member is the mind-spirit of Man, it follows that the structure and operation of its collectivity of lower members must similarly be mind-like, over all.

V. The universe has a positive value-bias.

Systems whose major components cooperate are stable or evolve Omega-ward; systems whose major parts conflict are unstable and break down toward Alpha. Since evolution is predominantly upward, the universe must have a positive value-bias. Unified Science's Natural Law is thus also Moral Law.

This, I submit, is a scientific revolution. It is a discontinuous way of regarding the problems of the physical, biological and psycho-socio-political sciences. It is a Copernican change of attitude, for it is not only moral but, in a profound sense, religious. For with its help we can consistently discern and, with unparalleled precision, constantly correct our relationship to the central order which, in the language of religion, is called the One or God.

5. APPLICATIONS OF UNIFIED SCIENCE: POLITICAL, ECONOMIC, ACADEMIC, AND RELIGIOUS

Over the past thirty-odd years, Unisci conversions have occurred in several ways. To acquaint you with this important change—the personality-change which transmutes a person or a group of people from the Lower to the Higher Industrial Period—I will report some of them here in the manner of a unified scientist with training in anthropology.[28]

Political

The first such transmutation of attitude occurred at Brooklyn College some twenty-five years ago. (I omit the person's name. She can, if she wants to, make it known herself.) Its background is as follows: my courses had been rather successful. Some of my students had formed a small club, the Systematic Social Science Club, to give them and me more time to discuss systematic social science, as we then called it, than there was in the regular classes, and to let them bring friends who were interested but who for various reasons could not take my courses. Our little club had, in sheer self-defense, turned the tables on the Marxist Club, the local Communist-front organization which had successfully infiltrated and was controlling many of the student societies on that campus.[29] All we had done was to clear their oratory of communications noise and fouling in the simple manner (as I later discovered) which Leibniz had advocated. And it had worked almost miraculously.

"If," Leibniz had written, "we could find characters or signs appropriate for expressing all our thoughts as definitely and as exactly as arithmetic expresses numbers or geometric analysis expresses lines, we could in all subjects *insofar as they are amenable to reasoning* accomplish what is done in Arithmetic and Geometry.... That would be an admirable help, even in political science and medicine, to steady and perfect reasoning.... For even while there will not be enough given circumstances to form an infallible judgement, we shall always be able to determine what is most probable on the data given. And that is all that reason can do." pp. 15–16.[21]

The Systematic Social Science Club had used my first model of Leibniz's General Characteristic very effectively. (It was simply the rectangular Cartesian coordinate system.) When one of the Marxist Club's best cadres came to our club meeting, as one usually did, and began to orate, one of our best cadres went to the blackboard, drew a coordinate system, and labeled it as follows:

Each time the Marxist discussed *exploitation*, a check mark went into quadrant 2: − for the workers, + for the employers (−, +). When he spoke of *expropriating the exploiters*, a check went into quadrant 4: + for the workers, − for the employers (+, −). If he

Unified Science's Moral Force

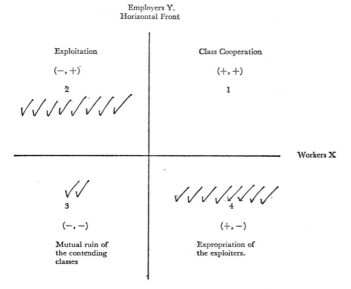

FIGURE V-3 Geometric mapping of the Marxist theory of human coactions. Transmutation of society from exploitation (quadrant 2) is affirmed to occur by "liquidation of the system's controller" (quadrant 3 or 4), and this to result in "classless" society.—The possibility of class cooperation (quadrant 1) is denied. Neither of these is possible in theory or in practice.

mentioned *the mutual ruin of the contending classes*, a mark went into quadrant 3: − for the workers, − for the employers (−, −).

Quadrant 2 always got the most checks. And we naturally agreed with the Marxist that these coactions between workers and employers are common and widespread. But then we asked him, or her, why quadrant 1 was empty; why she or he never mentioned anything that falls into quadrant 1.

"What's that?" they would demand.

"Plus for the workers, plus for the employers: (+, +)."

"But that's class-cooperation!" the Marxist would exclaim. "Marx and Lenin have said many times that there is no such thing as class cooperation.—Anyone who advocates it must be an agent of the Bourgeoisie!"

One of our people would then ask the Marxist where he would map the twenty-odd thousand profit-sharing companies that, we had learned in class, already then existed in the United States. (Today there are many more.)[30]

There would be, of course, division of opinion. Nonetheless the powerful performance of profit-sharing and multiple-management would begin to sink in.[31]

Economic

Another of my students would then describe Switzerland's Social Capital—the *vertical* front I had discovered in Switzerland.[32] A part of the Swiss public, the Majority, working cooperatively with the most brilliant and moral part of the Minority (management), were regulating an important part of the economy. Wherever a monopolistic company or association had forced prices so high as to be exploitive (−, +), *Der Regulator* (the Federation of Migros Cooperatives) built a factory, set up a chain of newspapers, organized a mass action, and so on, which forced the predators' prices down, so that they and the Swiss public became cooperative (+, +). This brilliant strategy—invented by Migros' founder, Gottlieb Duttweiler—has clearly redirected the Cold War in Switzerland: the monopolistic brand industries have been forced to lower prices, to improve quality, and to pay better wages and farm prices. In spite of bitter resistance they have been obliged to decrease their predation or even, in some cases, actually to transform it into cooperation with the Swiss public. Throughout large sections of the Swiss economy class struggle—predation and parasitism—is thus being changed by Migros into class cooperation.

As the student spoke, this social transmutation was mapped into a sequence of coordinate systems on the blackboard. According to a recent poll, Americans now consider Switzerland to be the world's best governed country. But practically no one understands the reason why. The reason is, that this country has undergone, and is still undergoing, an economic-political revolution. This almost invisible revolution has transformed large parts of Switzerland's economy from Capitalism into what they call *Social-Capitalism*. And this has made Switzerland much richer and pleasanter than it was before. For instead of being conducted by military, and thus destructive operations, it has been conducted by constructive economic-political-educational operations. This change occurred as follows:

Coordinate system 1 represents the traditional *Capitalist System*, as it existed in Switzerland in 1925 and still exists in varying degrees

Unified Science's Moral Force 193

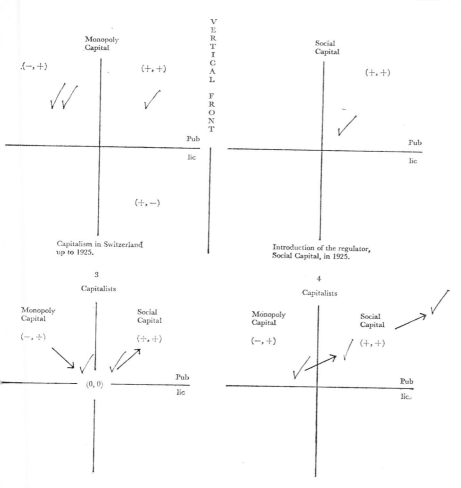

FIGURE V-4. Geometric mapping of the transmutation of an important part of Switzerland's economy from massive exploitation to massive class cooperation; from monopoly capitalism to an early form of Social Capitalism.

in nearly all countries of the partially Free World. Side by side, there are symbiotic $(+,+)$ industries and businesses, and monopolistic, predatory ones $(-,+)$; there are symbiotic labor organizations, and some who, by monopolistic union practices, systematically exploit and sometimes ruin management $(+,-)$.

Coordinate system 2 represents the formation of Switzerland's vertical front in 1925. "A group of far-sighted leaders, headed by

Gottlieb Duttweiler and supported today (1948) by one hundred and forty thousand common citizens—most of them with their families[33] —jointly created a region of cooperation in the following activities: Food distribution, industrial manufacturing, finance, farm production; press, movies, schools and book publishing; clothing, transport and tourist recreation, all this backed by a political movement especially strong in Zurich. These organizations have linked together enough of the spontaneous and scattered pockets of healthy Swiss resistance to both predation and parasitism to form a continuous cooperative front of both classes together, a vertical split from top to bottom of Swiss society. This is clearly shown by the fact that among the members of the Migros organizations and especially by the electors supporting its political movement (the Landring of the Independents) we find every class of the Swiss population, workers as well as manufacturers; producers as well as consumers; employees as well as employers; people of literary, artistic, and scientific professions as well as their directors, publishers, and administrators."[32]

Coordinate system 3 represents Switzerland's Social-Capitalist revolution; the long and continuing struggle between the two sides of her vertical front. By creating *mutually* beneficial stores and industries, Migros gives the Swiss public a *choice*, an *alternative* to the monopolists' exploitive industries, stores, and so forth. This choice transmutes a strategic volume of the monopolists' trade from predation (Group VI) to zero (Group 0).[34] Nobody is arrested or killed, no factories are destroyed. What people do is to transfer their trade, their economic ballots, from the Dominant Minority, the monopolistic exploiters, to their Creative Minority. The monopolists have mounted long, ferocious price wars, campaigns of vilification, and prosecutions in the courts. But the Swiss public has had the moral stamina and courage, and the intelligence to support their Creative Minority victoriously for nearly fifty years. They even forced down the prices of the international oil trust, and have now expanded their vertical front to defend their environment: Migros has declared war on the water polluting detergent manufacturers by giving the public equally good but non-polluting alternatives. The public is joining the fray enthusiastically.

Coordinate system 4 represents transmutation of the economy into Group IV (+, +), and possibly even into the Higher Industrial Period, the Human Period 7 (Figure IV-4). The ancient four temperaments (the modern "faces of Eve")—our inborn tendencies to *predation, parasitism,* and *mutual destruction* (or *withdrawal*) as well as to

Unified Science's Moral Force

cooperation—are built into each kingdom of the universe, including every person in it. Migros declares over and over that they rely on their competitors to keep them honest.

My conclusion is that Migros is perhaps one of the most highly developed cases of a vertical split in an industrial society. That it decreases class struggle, actually as well as theoretically, may be seen from the fact that Migros and the Landring are attacked furiously and tenaciously just by those, and all those, who only feel at home in, and know how to operate only in the ancient and fatal horizontal way: on one hand by all the Marxists—Communist and Social Democratic—and on the other by monopolistic big-business groups and their satellites in the middle class. Both the Marxists and the reactionaries know that as long as Migros and the Landring build cooperation vertically between large and healthy sectors of the two classes, and spread truth, kindliness, and humor, Switzerland cannot be disintegrated and incorporated either into the Communist or into the neofascist coalition. *On the contrary, these coalitions must be and are being disintegrated in Switzerland, and incorporated into the vertical structure.*

The Marxist student was overwhelmed. He had never heard of this moral alternative to the disasters of the Marxists' horizontal front; to their implacable warfare between Majority and Minority by which the Communist Party, the New Class, wrests control from the Dominant Minority in country after country and destroys most of the Creative Minority along with it. The Marxists simply had nothing to say. And none of them ever came back to a meeting of our club which, incidentally, came to be called the *Plus, Plus Club*.

Presently we heard through the grapevine that there was an uproar in the Marxist Club. Its Communist organizer was furious. "I send you there to take them over," he had yelled, "and you come back and argue with *me*!"—So he reversed his tactics and hung up an Iron Curtain. His club members were forbidden ever to come to our meetings; even to talk to us privately.

This did not satisfy his best cadres.—Not being allowed to discuss means, they saw clearly, that many Communist arguments can't survive geometric mapping.—Many of them dropped out of the Marxist Club, and soon our Plus, Plus Club was organizing joint meetings with the Philosophy Club, Psychology Club, and Sociology Club, to hear constructive speakers such as, for instance, Erich Fromm.

When, later, I read Leibniz' prediction of the moral force of his Universal Characteristic, I realized that he had known what he was talking about, and that his prediction had been fulfilled at Brooklyn College. Even our first, Cartesian model of his General Characteristic had redirected the Cold War on the Brooklyn campus: it had damped the conflict, fostered by the Marxist Club's negative value-bias, changing most of that conflict into the cooperation furthered by the Plus, Plus Club's positive value-bias. In due course, this experience, and others like it, led Harold Cassidy and me to write, and privately distribute, *Plain Truth—And Redirection of the Cold War*.[35]

With this as background, I can now hope to convey some of the meaning of the following experience: One afternoon at Brooklyn College, when everyone had left the classroom and I was picking up my books and papers, one of the students, a quite pretty girl, came back. Closing the door, she asked whether she could speak with me.

"Mr Haskell," she burst out, "I don't know what to do! The whole world has changed!—I didn't have any hope before; but now I have hope, even confidence! Now I know that the world has a future! I want to help you!—I'll do anything: scrub your floors, type your manuscripts—anything!" Tears were running down her beaming face.

I knew her only as a student who chose to sit in a back row and hardly ever spoke in class. Her papers had never struck me as exceptional. Yet here, suddenly and without warning, she was showing me the deepest understanding of them all! She had undergone even more than a Copernican change of understanding.—I asked her to sit down and let me think about what she had said.

After a while I told her that I understood what she was talking about. When I was a student, I had despaired far more, perhaps, than she. But unification of the sciences, in my mind, had opened for me a road to life, to the future. It was this opening which gave her and many of the other students hope. If she wanted to help me, the best way I could think of would be for her to join the Systematic Social Science Club, (the Plus Plus Club), and help it to organize Brooklyn College.

She dried her face, smiled, thanked me, and said she would do it. We shook hands.

Next day I inquired about her among the students. They said that she had been extraordinarily affected, and had cried several times after class. They were glad that she had talked to me, and that she now had something concrete to do. Here, I realized, was a remarkable event, a religious experience like some of those described by William James.[36] I had experienced and described "The Religious Force of Unified Science" long before.[11] Now I had seen it grip somebody else.

Within a few days this girl and her friends had collected, among the college's students, 250 signatures on a petition requesting of the college's Administration—instead of just modified sociology and anthropology courses—a completely new kind of course, Unified Science.

Naturally, her petition seemed fantastic to the administration, and was rejected. But it put in train a chain of events which, I see now, forced me against my will onto a far more feasible path. I was forced by left-wing colleagues out of the academic world for many years. But the course of action on which I then set forth may possibly result in a fast enough spread of Unified Science for Man's Empire to survive the catastrophes which its Lower Industrial Culture can no longer possibly avoid.

Academic

Unknown to myself, I presently "repeated" (on a higher level) one of the main steps by which leading 17th century scientists transmuted their obsolete medieval universities into the new kind of institution called for by the emerging Lower Industrial Culture: For many years in the mid-1600s, a group of the new (Copernican, Baconian and Galilean) scientists met in London, in and about Gresham College. They usually met after the lecture of the Professor of Astronomy, either in his rooms or at a nearby inn. Some people called them, rather aptly, The Invisible College.[37] In time, however, when the British Monarchy was restored and the new King needed to make a new departure, he conferred upon this group—in which the original Copernican change was most highly developed—a Royal Charter, transforming it into the Royal Society.[38] And the Royal Society—along with its sister societies all over Europe and America—transformed the slowly yet centripetally evolving medieval university into the centrifugal modern multiversity, capable of making enormous disintegrative progress (Figures IV-9, 10).

"It was indeed a new departure," observes Arthur Koestler. "The range and power of the main sense organ of *homo sapiens* had suddenly started to grow in leaps to thirty times, a hundred times, a thousand times its natural capacity. Parallel leaps and bounds in the range of other organs were soon to transform the species into a race of giants in power—without enlarging its moral power by an inch. It was a monstrously one-sided mutation—as if moles were growing the size of whales, but retaining the instincts of moles. The makers of the (first) scientific revolution were individuals who in the transformation of the race played the part of the mutating genes. Such genes are *ipso facto* unbalanced and unstable. The personalities of these 'mutants' already foreshadowed the discrepancy in the next development of man: the intellectual giants of the (first) scientific revolution were moral dwarfs." p. 352.[20]

The present book has shown, I hope, that Koestler's diagnosis, while basically valid, is technically somewhat off the mark. (If it were not, no human effort could hope to be of any use at all.) The main defect of all the Lower Industrial peoples—except, perhaps, Switzerland's—is the same weakness in determining moral direction which Toynbee found to characterize disintegrating Literate civilizations; a defect whose symptoms he has called *promiscuity, abandon, truancy,* and *drift*.[16] The difference is not basically one of physical size—of moles and whales, of dwarfs and giants. It is, rather, one of speed and power: say Roman triremes versus modern hydrofoils or jets. Lack of a compass can make triremes drift or move promiscuously; but it transforms hydrofoils or jets into roaring, searing balls of flame.—And, while Koestler senses correctly that mutations are at work in transmuting Literate into Lower Industrial civilization quantitatively (Figures IV-2 and IV-4), the failure of Industrial civilization to transmute qualitatively into Group IV (+, +), where alone it can survive, is probably not related to genetics. It is due to the order in which intellectual transmutations necessarily come about:

The Periodic Law is so constructed that only after vast knowledge of some given kind of system's properties (say the atoms') has been amassed, can it be seen that they are a function of *anything*, let alone of such an elusive, subtle thing as coaction, the moral relation between the system's work component and controller.—Because of this necessary sequence of events, even the finest human beings—which almost all the scientists I know most surely are—have first to become "giants" of cognitive intellect (Lower Industrialists) while long remaining moral "dwarfs"; that is, without *scientific* moral orientation.

It was Mendeleev's completion of the Periodic Law a century ago in chemistry (and only Mendeleev intuited its cosmic or philosophical implications) which prematurely organized the physical sciences. And it will be the extension of this organization to the social and biological sciences which will complete this basically moral orientation, transmuting Koestler's "moral dwarfs" into (computer-assisted) moral giants, and giving our society's controller moral direction, and thus balance and stability.

Today, in the Seventies, hindsight reveals these things. But back in the Forties, only a few colleagues and a handful of students had the faintest notion of what had to be done. I therefore instinctively organized the first Invisible College for our time: the Council for Unified Research and Education, Inc. was founded during the Centenary celebration of the American Association for the Advancement of Science.

And what was its objective, from its beginning in 1948? We announced it visibly enough in *Science:* "An OSRD-like organization which may hope to succeed in advancing social (and biological) science through the stages of natural classification (J. S. Mill) and evolution theory (von Bertallanffy and Norbert Wiener), into that achieved by the physical sciences, where scientific fields are connected, and science is closely linked to philosophy and technology. The possibility of such coordination emerged with the independent discoveries of parts of the same conceptual scheme by students of plant, animal, and human coactions (as Leibniz had predicted) ... Should this scheme prove to be a natural classification (J. S. Mill), it would create conditions for rapid coordination and advance of social science, as the Periodic Table did in chemistry."[39,40]

The announcement ended with a prediction that only a few Brooklyn College students could at that time begin to understand: "It is anticipated that the philosophical, scientific, and technological structures of Western and Eastern ideologies about conflict and cooperation will have been sufficiently clarified by then to make possible their gradual displacement by an advanced social science which is systematic, useful, and universally accepted, as physical science already is today."[40]

A few years later, I discovered that what C.U.R.E. had undertaken was, actually, to carry out the Royal Society's second but necessarily unfulfilled objective: in 1663 its Secretary, Robert Hooke, had announced the Royal Society's objectives in two parts: the first part was empirical and inductive: to discover the properties of all the natural kinds of systems. That is, to discover and describe the myriad parts of the universe. This first objective the Royal Society and its sister societies around the world have now carried out successfully enough: it has resulted in the multiversity, Figure IV-9. The second part of its program was deductive and theoretical: "The compiling of a complete system of solid philosophy."[38] That is, to assemble this immense warehouse of heterogeneous parts into a coherent, working, viable whole: *Unified Science*.

By 1954 it had become clear that C.U.R.E. had undertaken a far more difficult task than we had realized in 1948. On Harold Cassidy's advice, we then became "invisible": C.U.R.E. suspended its meetings in order to concentrate on the necessarily mental-spiritual solution of its data-theory assembly problem. We worked on it "invisibly"—as individuals and pairs and small groups—for fifteen years. And in 1969, the hundredth anniversary of Mendeleev's Periodic Table, again at a meeting of the American Association for the Advancement of Science, we briefly summarized the results of our invisible work. You see them outlined in this book.[41]

What we have done for science as a whole is at best only the sort of thing that Mendeleev did for chemistry. We have designed a model and some parts of a machine-tool for this model's manufacture. But fortunately we have done this much in time: In 1970, in his article on "The Club of Rome and the Predicament of Mankind," the Club's president, A. Peccei, published a call for a systematic viewpoint; a project which it takes *decades* to execute.

"The Club of Rome," says the *Science Policy Bulletin*, "comprises some 50 scientists, planners, intellectuals and industrialists from Asia, Africa, Western Europe, and North and Latin America, and is concerned with global problems of the techno-scientific (Lower Industrial) society. The 'action oriented' Club of Rome 'believes it is still possible . . . to meet this unprecedented tangle of problems before it outstrips our capacity for control.' The Club's first objective is 'to acquire and spread an in-depth understanding of the present critical state of human affairs and of the narrowing and uncertain perspectives and options which are likely for the future, if present trends are not corrected. The second objective is, then, to recognize and propose new policy guidelines and patterns of action capable of redressing the situation and keeping it under control.'"[42] Then comes the call for the leading link, the strategic factor: Peccei affirms that "the Club feels there is an urgent necessity for a 'Copernican change' in attitude, to shift from a fragmented viewpoint to a systematic viewpoint."[19]

The book you are reading and its waiting line of sequels display this change of attitude, together with the concepts and metalanguage necessary for transmitting it. Their successful introduction into the multiversity will transform it in the manner foreseen and predicted in Figure IV-11, the New University. This revolutionized institution's alumni—organized specialists and generalists, Figure IV-12—will spread this New-Copernican attitude from kindergarten through primary, secondary, and tertiary schools, to graduate schools.

Correct and successful examination, diagnosis, prescription and prognosis are essential. But while the Club of Rome, our multiversities and the Scientific Community tend to stop there, the leadership procedure-sequence goes on, inexorably: after *prognosis* must come *execution* of the prescription. And this can be done only by the system's work component, the Majority, which Communists call the *masses*, and Swiss Social Capitalists call the *general public*. The Club of Rome wants to *instruct* the public. But to execute its prescriptions the public must be *led*! And finally, after each action, must come the feedback operation, retrospection: careful comparison of the execution's outcome with the *prognosis*, so as to correct and improve the system's later performances. Only retrospection closes the system's circuit and guarantees morally oriented development.

Theory becomes a social force only when it has gripped the masses, Marx declared. But the masses have, by definition, 1, 2, 3 and 4 human levels of abstraction, while theory includes levels 5, 6, and 7 (see Figures IV-1 to 4). To grip the masses, theory has to be transposed to every level. For without transposition, theory cannot grip the masses and is sterile. And who knows how to transpose theory better than organized specialists and generalists?

It is not, however, primarily scientists, but leaders of the great public who can execute prescriptions, *carry out* the patterns of action now being formulated. Their training is provided for in the New University's School of Management, shown on the left-hand side of Figure IV-9.

Yet plenty of our colleagues who cannot change are apt to drag their heels: some of the best established, most reputable ones refused even to *look* through Galileo's telescope or Pasteur's microscope! Their fate is to be dragged, kicking and screaming, in the dust of the chariot of becoming.

The men and women who will take upon themselves the chariots' heavy harness, and change the death-ward course of the world's development are, first of all, young people, and in all cases positively oriented ordinary citizens of vision and faith.

How, then, can unified scientists and engineers, who must work out the diagnoses and prescriptions, move the Majority to their correct *execution?* By learning how these so-called common people think, talk, and act. And that means living among the people, the Literate and sub-Literate Cultures, and learning how to translate scientese into, and out of, their mythic or religious language without changing Unified Science's basic meaning.

Religious

What are religions? They are the unified sciences of pre-industrial peoples; the simpler peoples' coherent understandings of the world, in both its local and its universal aspects. That is what anthropologists mean when they say that primitive societies are sacred societies.[27] If this is so, its converse follows logically: Unified Science is the religion of the industrial peoples, scientific culture. It is the modern updated form of religion called for by John A. T. Robinson and his theological colleagues around the world. If "in the final analysis our compass must be our relationship with a central order",[4] and if Unisci provides this compass for the scientific portion of our global industrial society, then it fulfills for it the basic role which religion—in its diverse, increasingly sophisticated forms—has played in all the human Periods' literate and sub-literate cultures.

The great religious and artistic leaders of mankind have relied on, but stopped with, inspirations, revelations, and flashes of insight. They have stopped where creative scientists *begin*. What gives our scientists their immense authority, which has outstripped the authority of traditional priests and preachers, is their conscientious, detailed empirical verification of their revelations. This has, within each separate field, consistently winnowed out false revelations, corrected half-false inspirations, completed incomplete insights. The consequent structural correspondence of their thought with reality has conferred dominion upon modern technologists; the power to solve individual problems of health, transport, communication and so forth, at which men of traditional religion boggle. For unification of the corrected and verified separate scientific revelations now results in the moral orientation of the powerful scientific effort as a whole.

With moral orientation, science has come full circle: When unified, as here, science's revelations no longer conflict with, but reinforce and expand the most important revelations of the great religions. For the value-bias of both cultures, humanistic and scientific, has now become clearly positive. A host of ancient sayings, revelations, and myths have been, and can now be mapped geometrically, and their values assessed in the unequivocal idiom of science. Conversely, large and increasing bodies of scientific findings, and of the practices they generate, can be translated into effective religious and political idioms.

Unlike the detectors of the one-field scientific specialists, Figure IV-10, which screen out and reject most of religion's doctrines, and all its values, the detector of Unified Science, Figure VI-12, accepts the great majority of them. And more than that, it extends some, corrects some, completes some, and profoundly illuminates the whole, as the traditional religious attitude in turn illuminates, for many, Unified Science as a whole. The semi-conscious process which Toynbee groups under the general term *mimesis* herewith becomes the most important of technologies.

Joseph Campbell, author of the *Masks of God*, points out that though not true in a literal sense, a myth is not what it is considered to be in everyday speech—a fantasy or misstatement.[43] *It is rather a veiled explanation of the truth.* "We have seen what has happened to primitive communities unsettled by the white man's civilization," he observes. "With their old taboos discredited, they immediately go to pieces, disintegrate, and become resorts of vice and disease. Today the same thing is happening to us."

Gerald Clarke's comment in a *Time Essay* has the sharpest point: "The mythologists (such as Joseph Campbell) are not providing myths, but they are indicating that something is missing without them. They are telling modern man that he has not outgrown mythology and will never outgrow it".[44]

The thing that has been missing is a reliable, practical, and spacious bridge between our old and enormous Literary Culture, whose language is basically myth in the sense of veiled truth; and our new Scientific Culture whose language is basically technical—in the sense of, for many, incomprehensible truth. I shall now bring evidence to show that with this bridge, the missing "something" is in place; that now at last science has come full circle. That the people who belong to our Two Cultures can now communicate across the Industrial world's cultural chasm, can orient each other, and together can transmute our Lower Industrial civilization as a whole into the Higher Industrial Period, Human Period 7. Switzerland has transmuted herself spontaneously. She is our little pilot plant, presenting us with decades of research and development, giving us vast amounts of data and experience on which to base our Creative Centrist alternative to the political ideologies of Extreme Left and Extreme Right (Figure II-16) on how to damp out their fatal rhythm of disintegration.

To illustrate this bridge's operation, I take some typical passages from the *Autobiography of Harry Emerson Fosdick*, a former pastor of my church, the Riverside Church in New York City.

"Mankind," wrote Dr. Fosdick, "desperately needs what Christianity *at its best* has to offer—that idea has become ever more urgent and commanding. I emphasize *at its best* because Christianity can be and often is perverted, corrupted, degraded, until far from serving good ends it becomes a deplorable evil. I grow weary at times with preachers who, without clarifying definitions, set over against each other words like 'Christianity' and 'secularism,' as though secularism were cursing the world and Christianity alone could save it. The fact is that so-called 'Christianity' at its worst has produced some of the most hideous persecutions, wars and fanaticisms in history, and that today it is sometimes bigoted, superstitious, intolerant, socially disruptive, while so-called 'secularism' is sometimes humane, ethically-minded and socially constructive." pp. 267–8.[45]

Dr. Fosdick's sermons and books display hundreds of concrete examples of Christianity and secularism at their best and also at their worst. These examples can now be mapped into the Periodic coordinate system, as in Figures V-3 and 4, and thus be given more clarifying definitions than humanists have here-to-fore commanded.

"I do not wish to use the word 'Christianity' as though it were an unambiguous term," he wrote a little further on. "One needs to define what one means by it." He then uses what scientists call a *type specimen* or a *paradigmatic case*:[46] "For me the essence of Christianity is incarnate in the personality of the Master ... and in the fundamental principles of life's conduct which Jesus of Nazareth exhibited. I am sure that the world today desperately needs his faith and his way of life, and that without them there is no hope."[45,47] Each concrete aspect of this paradigmatic case can be mapped unambiguously into the Periodic coordinate system, and the word 'Christianity' thus rendered quite unambiguous.—A number of its parts will, I predict, be found to belong to the Literate Culture (Period 5), and to require *transposition* in order to regain their original meaning for Industrial culture. And others of its parts will have to be completed.

"This conviction [that we must emulate this paradigm] has been forced home on our generation by our disillusionment with some of the reliances in which we trusted for the salvation of the world—science and education, for example ... They are only instruments and the crucial question on which everything in the end depends is what kind of people—with what undergirding conviction about God and man, with what quality of character and with what ethical standards—are going to use them." (*Ibid.*)

How relieved Dr. Fosdick would have been had he witnessed the moral orientation of the sciences resulting from their unification may be gathered two paragraphs later on: "Science (that is, the one-field disciplines) has brought mankind proximity, the ends of the earth woven together in intercommunication and interdependence, but it cannot provide the ethical quality which, transforming proximity into fraternal community, saves proximity from becoming tragedy." (*Ibid.*)

"I have lived into a generation," Dr. Fosdick continued, "where not science alone but education too 'has created a world in which Christianity is an imperative.'" He then lists almost all the defects which the crisis ridden university's now perfectly feasible assembly plant, Figure IV-11, has been shown to correct: "Facts without values, fragmentary specialties with no integrating philosophy of life as a whole, data with no ethical standards for their use, techniques with no convictions about life's ultimate meaning or with corrupting convictions—here, too, a panacea has turned out to be a problem. What quality of faith and character is going to use our educated minds?"p.271.

Answer: The quality wherein Unified Science and religion at its best coincide; and that is probably throughout.

The other major area in which Fosdick's religion and Unified Science reinforce each other he classes in his autobiography's index under *Counseling, personal:* "People come to church on Sunday with every kind of personal difficulty and problem flesh is heir to. A sermon [or, I presume to interject, a Unified Science lecture] was meant to meet such needs; it should be personal counseling on a group scale.... That was the place to start—with the real problems of the people. That was a sermon's speciality, which made it a sermon, not an essay or a lecture." Then Dr. Fosdick defines for his church services much the same objective which I had defined for each of my classes: "Every sermon should have for its main business the head-on constructive meeting of some problem which was puzzling minds, burdening consciences, distracting lives, and no sermon (or class) which so met a real human difficulty, with light to throw on it and help to win a victory over it, could possibly be futile." pp. 94.[45]

Now let us turn to concrete cases other than those at Brooklyn College back in the 'forties. Has Unified Science increased its capacity not just to instruct, but to inspire people to creative *action*?

One morning in February 1971, as I was working in my study here in New York, one of my students, an Englishman named John Harries, called me by telephone from New Haven, Connecticut. I had just started another seminar in Unified Science at Jere Clark's Interdisciplinary Center, Southern Connecticut State College.— Harries informed me that he was offering me a room free of charge, so I would not have to stay at motels after my evening seminars·

When I boggled, he explained that he was director of a group of students and young working people called the Unification Church. He said they had had a meeting and had voted to offer me a room in their house, free of charge.

I was, of course, delighted. But I said I would certainly want to pay my share of the house rent.

He answered that this would not be necessary: they wanted to subsidize my work!

When I protested that they could hardly have much of an idea of my work, since the course had just started and he had attended just one session, he laughed and said that if it would make me feel better, I could contribute whatever I saw fit.

But when, after the next seminar, I arrived at their house, they surrounded me with warmest hospitality. You would have thought I was their dearest relative!—They all slept on the floor—men, of course, strictly separate from women. But for me they brought a four-post bed from the cellar and set it up in a large private room. Each night, when the time came for their evening prayer session (at about 10:30), one of then took my suitcase and escorted me upstairs to my room. Thus was established, without a spoken word, a geographic separation between my uncompromisingly scientific classes and their uncompromisingly religious prayers. Quite possible embarrassments were thus forestalled.

But in due course I happened to pick up a copy of their *Divine Principle* and in my next seminar I translated some of its religious principles into scientese.[23] Their *horizontal* and *vertical* principles seem to assemble into the Periodic coordinate system; their *object* and *subject* can then be interpreted as Work Component X and Controller Y; their *give-and-take* then becomes *symbiosis* $(+,+)$, extending their vocabulary to all coactions, and so forth. Not everything translates, of course: from science's point of view, a lot of what they say has got to be regarded as symbolic. But they seem able to accept that. This leaves us smilingly convinced that we are, in our very different scientific and humanistic ways, describing the same thing.

A few weeks later I found myself at their United States headquarters in Washington, D.C., giving a lecture on *The Intertranslatability of Unified Science and Unified Religion* amid exclamations of "That's wonderful! That's absolutely perfect!"

I had, of course, to keep replying that it is most certainly not perfect, but that it may be perfectible.—That is a service which scientists must constantly and firmly render to men of religion, to keep our minds open to correction and to growth. What deadlier paralysis is there than the illusion of having reached perfection?

For the next term at Southern Connecticut State College, its 1971 summer session, the Unification Church sent students to New Haven from Washington, Philadelphia, New York, and a physicist and a political scientist across the continent from California. We lived together at the New Haven Center. The physicist, Glenn Strait, conducted two evening sessions there each week to coach and brief all those who needed it. Next morning at eight o'clock, unresolved questions were taken up in class. "Counseling, personal" sometimes went on by appointment, and sometimes by students' impulsively coming to my room.

By midterm, the class had divided itself along the Two-Culture line into Scientists and Humanists. I designed the examinations so that one set of specialized questions was directed to scientists, a different set to humanists. But all basic questions belong to both cultures. They were directed to, and answered by, everyone.

The class's extensive term papers, written along these clearly converging lines, turned out so splendidly that I suggested the possibility of publishing them as a book. The students enthusiastically elected an editorial board, and when the book is ready, and its title decided upon, we will submit it to a publisher. Have not the Two Cultures come together, as C. P. Snow predicted, in the United States?

Early in 1972, the founder of the Unification Church arrived in the United States. Sun Myung Moon is a South Korean philosopher, raised as a Christian and trained in electrical engineering in Japan. His Church's half million profoundly dedicated members are citizens of some twenty-six countries in Asia, America, and Europe.

At our first meeting, in which Mr. Moon was flanked by three Korean interpreters, and I by the directors of two of his American centers, he announced that he wished me to organize an international conference so that the world could become acquainted with Unified Science.

I was, of course, elated. This would transform our invisible college into the visible executor of the fatally unfinished part of the Royal Society's program of 1663. Mr. Moon proposed that six eminent scientists be invited from Europe, four from Asia, and ten from America. He suggested that we invite fifty or sixty observers, and that C.U.R.E's membership be expanded to all the continents.

I replied that if no strings were attached—if C.U.R.E. could decide all questions of subject, persons, discussions, and so forth without interference from his Church of any sort, C.U.R.E., Inc. would gladly accept this offer.

Mr. Moon assured me that there would be no interference on the part of the Unification Church, just as there had been none with my courses at Southern Connecticut State College. He concluded by asking me to draw up a budget—including aeroplane tickets, a first class hotel, and all the rest—and to submit it to him as soon as possible.

The conference on *Moral Orientation of the Sciences* was thus scheduled to follow shortly the publication of our invisible college's Copernican change of attitude embodied in this book.

This publication is of course indispensable. But by itself it would certainly be too little. To paraphrase Harry Emerson Fosdick[45] p. 99, a unified scientist's business is not merely to discuss moral orientation of the sciences, but to persuade scientists to orient themselves morally; not merely to debate the meaning and possibility of Unified Science, but to produce this moral synthesis in the lives of his listeners; not merely to talk about the available moral force of Unified Science to bring victory over academic disintegration and ecotechnical temptation, but to send people out from his classes and conferences with victory in their possession. A unified scientist's task is to create in his associates, fellow scientists and students the thing he is talking about. (End of paraphasis.)

One logical place to practice this is, of course, our Riverside Church where Dr. Fosdick preached. In December, 1971, I wrote its present Preaching Minister the following New Year's letter:

Dear Ernest Campbell, best wishes for the coming year! This year will mark the fusion of our ancient Literary Culture with our new Scientific Culture: and thus of religion with science. It was bound to happen: the deeper meaning of such symbols as A and Ω was bound to appear as sciences merge and illuminate each other. The Bible passages say "... the beginning and the end ..."; now Unified Science completes these sentences and says *of what*.

Since I am a member of your Church, Dr. Campbell, you will probably be involved in these developments. People will ask you what your views are on me, as they have asked me how I can listen to you preach.—I've listened many times, so I can answer; I have, in fact, listened to preachers all my life—my father, grandfather, and (in print) great grandfathers among them.—What I have done is, to make it possible for conscientious scientists and their millions of students to listen to men of religion as they have *not* done, now, for decades. I have also made it possible for you to listen to them, as you have *never* really done.

Not actually, of course, to *them*; but to those among them who—reviewing their one-field, and thus amoral, specializations—participate in this unification of sciences, in which most of our ancient values reappear in modern form.

A dialogue now becomes possible: Unified Science appears to be what Teilhard de Chardin calls *depersonalized* religion; religion which can thus be invested with personality, such as that of Jesus. (That is what *in-carnation* now comes to mean.) But as mankind approaches and reaches our planet's saturation point, incarnation and mankind's ways must change accordingly. If they do not, implacable judgement, which transgression of moral law brings on, will surely and certainly be executed; and that, by Mankind upon itself.—Unified Science corroborates many ancient, intuited predictions, terribly.—It also shows how to avoid their happening.

What, Dr. Campbell, are you going to do?

Edward Haskell

P.S. I'm xeroxing this letter and will include it in the concluding chapter of our book, nailing my theses to the cathedral gate.

To this, Dr. Campbell graciously replied:

"What I propose to do is first to congratulate you for seeing this venture through. Second, I intend to read the book insofar as my limited scientific perceptions will permit me to do. Third, I will give you my honest response to the book, something I frequently do in line of duty under the general heading of book reviews."—Our dialogue has thus begun.

The crucial dialogues, however, will be among scientists, for our first objective has to be moral orientation of the sciences. Only with its attainment, does our ultimate aim become even possibly attainable: moral orientation of the Earth.—That is the way it's said in uni-scientese. In the ideolect of our religion this goal is called the Kingdom of God.

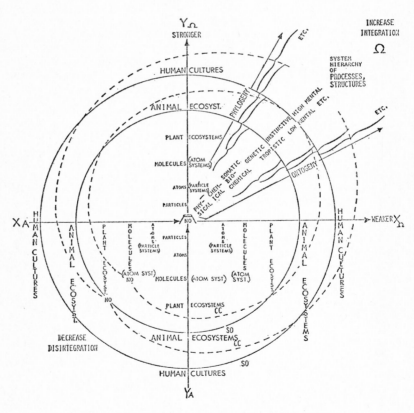

FIGURE V-5 The Central Order (1972 model).

Whatever its name. this objective will probably be approached—as far as it depends on Man—to the degree that our sciences (including political science), our philosophies (examples of which are listed in the fold-out chart), the arts, technology and religion come to apply the coaction compass to their developments.

NOTES AND REFERENCES

1. This chapter was written at the suggestion of the publisher's editor, Dr. Ervin Laszlo, Professor of Philosophy at the State University College, Geneseo, New York, to whose many constructive proposals this book owes its final structure.
2. Jonas, Hans, *The Phenomenon of Life—Toward a Philosophy of Biology*, Bell Publ. Co., New York, 1968.
3. This diagram, based upon *Two Modes of Thought* by James B. Conant is taken from my xeroxed book *Unified Science—Assembly of the Sciences into A Single Discipline*, Volume I, *Scientia Generalis*, Chap. 7. It has been used as a textbook at the New School for Social Research in New York and Southern Connecticut State College in New Haven, and is being readied for publication. (See, Conant, James, B., *Two Modes of Thought*, Trident Press, New York, 1964. Haskell, Edward, *Unified Science—Assembly of the Sciences into a Single Discipline*, Vol. I, *Scientia Generalis*. Preface and a chapter by Harold G. Cassidy. Offset-printed by N.I.H. 1968; xeroxed by IBM Systems Res. Inst., New York, 1969.)
4. Heisenberg, Werner, *Physics and Beyond*, Harper & Row, New York, 1971.
5. See, for example, *The Psychology of Invention in the Mathematical Field* by Jacques Hadamard. Or *Reason and Chance in Scientific Discovery* by R. Taton. (See Hadamard, Jacques, *The Psychology of Invention in the Mathematical Field*. Princeton Univ. Press, 1945. Taton, R., *Reason and Chance in Scientific Discovery*, Philosophical Library, New York, 1957.)
6. The term *ectropy* was, I repeat, coined by W. V. Quine in 1969, replacing such inelegant terms as *negentropy* and *negative entropy*.
7. Lodge, George T., "Measurement of Man-Machine System Performance," *Proceedings of the XVIIth International Congress of applied Psychology*, Liege, Belgium, 1971.
8. Toffler, Alvin, *Future Shock*, Bantam Books, New York, 1970.
9. Lippmann, Walter, *The Public Philosophy—On the Decline and Revival of Western Democracy*. Little, Brown, Boston, 1955.
10. The concept, *paradigm*, both in Figure IV-10 and IV-12 is due to Thomas L. Kuhn's *Structure of Scientific Revolutions*. In the case of *public philosophy*, the concept is generalized to non-scientific ideologies. (See Kuhn, Thomas, *The Structure of Scientific Revolutions*, First ed. University of Chicago Press, 1962.)
11. Haskell, Edward F., "The Religious Force of Unified Science," *Scientific Monthly*, June, 1942, Vol. LIV, pp. 545–551.
12. ———, "Unified Science—The Public Philosophy of the Space Age," *Connecticut Review*, Hartford, Conn., April, 1969.

13. "The Public Philosophy of the Space Age," a chapter in a symposium volume. Gordon and Breach, Science Publishers, New York, 1972.
14. Bertalanffy, Ludwig von, *General Systems Theory*—Foundations, Development, Applications. Braziller, New York, 1968.
15. Toynbee, Arnold J., *An Historian's View of Religion*. Oxford Univ. Press, 1956.
16. ———, *A Study of History*, Abridgement of Vols. I-VI by D. C. Somervell. Oxford Univ. Press, New York, 1947.
17. McNeill, William., *The Rise of the West*—A History of the Human Community. Univ. of Chicago Press, 1963.
18. The present concept and definition of *System-hierarchy* were formulated in discussions among Cassidy, Quine and Haskell between 1964 and 1969.
19. Peccei, A. "The Club of Rome and the Predicament of Mankind" in *Science Policy News*, September 1970 (pp. 13–14).
20. As a matter of fact, it was obvious to at least one man: Aristarchus of Samos not only discovered the Earth's rotations some twenty-odd centuries before Copernicus, but even the elliptic shape of its orbit, which Copernicus never did discover. Copernicus, moreover, had heard about Aristarchus. (See Koestler, Arthur, *The Sleepwalkers*—A History of Man's Changing Vision of the Universe. Macmillan, New York, 1959.)
21. This fact cannot be judged or even seen from the perspective of any one-field specialist: it seems to belong to anthropology, but it involves higher levels of abstraction. These include anthropology itself as a datum; also modern religion, philosophy, science, and ideology in general. This inclusiveness was predicted by Leibniz for the Universal Characteristic, of which the Periodic coordinate system is a model. (See *Leibniz Selections*, Phillip P. Wiener, editor. Scribner's, New York, 1951.)
22. Robinson, John A. T., *Honest to God*, Westminster Press, Philadelphia, 1963.
23. Kim, Young Oon, *Divine Principle and its Application*. Holy Spirit Association for the Unification of World Christianity. 1611 Upshur St., N.W. Washington, D.C. 20011, 1968.
24. Edwards, David L., editor, *The Honest to God Debate*. Westminster Press, Philadelphia, 1963.
25. Snow, C. P., *The Two Cultures*—And a Second Look. New American Library, New York, 1963.
26. I am scheduled to teach a course at Southern Connecticut State College in 1972 titled, *Unification of the Two Cultures: Scientific and Literary*.
27. Durkheim, Emile, *Les formes elementaires de la vie religieuse*, F. Alcon, Paris, 1912.
28. The writer is an Associate of *Current Anthropology* and Fellow of the Society for Applied Anthropology,
29. The writer knows something about Communist organization: He had been a member of the Communist party from 1933 to 1936, and had been president of a large Communist-front organization. He had chaired a delegation which had travelled to and worked with Communist organizations in ten countries, including the Soviet Union. During this trip he had been arrested three times. He had spent an afternoon in the home of George Dimitroff, General Secretary of the Communist International (near Moscow), discussing (in Bulgarian) the world-program of the Comintern. Though he had dropped out of the Communist Party in 1936, he was

awarded the decoration "9th of September" by mail "For your social activity as a friend of Bulgaria."

30. Jehring, J. J., *Profit Sharing: The Capitalistic Challenge*. Profit Sharing Research Foundation, 1718 Sherman Ave., Evanston, Ill., ca. 1965.
31. McCormic, Charles P., *The Power of People—Multiple Management Up to Date*. Harper, New York, 1949.
32. Haskell, Edward F., "Switzerland's Vertical Front—The Migros Federation of Cooperatives in the Light of Systematic Social Science." A chapter in *Gottlieb Duttweiler, dem Sechzigjährigen, Dank und Kritik*. Speer-Verlag, Zurich, 1948.
33. In 1971 Migros had 909,602 registered members in a nation of about 6 million.
34. The nearly one million registered Migros members have several times as many fellow-travellers, who fight beside them in the Social-Capital trade wars and anti-pollution trade wars.
35. Haskell, Edward F., and Harold G. Cassidy, *Plain Truth—And Redirection of the Cold War*. Privately printed and distributed, 1961.
36. James, William, *The Varieties of Religious Experience*. Longmans, Green, New York, 1902.
37. "The Invisible College," a chapter in *Roots of Scientific Thought—A Cultural Perspective*. P. Wiener and A. Noland, eds. Basic Books, New York, 1957.
38. Lyon, A., *History of The Royal Society*. Cambridge Univ. Press.
39. The Office of Scientific Research and Development organized the United States' scientific effort to win World War II. (See Irving Stewart, *Organizing Scientific Research for War*, Atlantic, Little Brown, Boston, 1948.) C.U.R.E. is now organizing the world's scientific effort to win World Peace I. (See below.)
40. *Science*, "Symposium on Cooperation and Conflict Among Living Organisms." Sept. 3, 1948 (pp. 263-4).
41. *Program, American Association for the Advancement of Science*, Boston, Mass., Dec. 1969 (p. 248).
42. *Science Policy Bulletin*, October 1970 (p. 2).
43. Campbell, Joseph, *The Masks of God* (4 vols.). Viking/Compass, New York, last vol., 1968.
44. Clarke, Gerald, "The Need for New Myths," *Time* Magazine, Jan. 17, 1972 (p. 50).
45. Fosdick, Harry Emerson, *Autobiography of Harry Emerson Fosdick—The Living of these Days*. Harper, New York, 1956.
46. Quine, Willard van Orman, *Ontological Relativity and Other Essays*. Columbia Univ. Press, 1969.
47. Compare to this the following statement by Werner Heisenberg: "If we ask Western man what is good and what is evil, what is worth striving for and what is to be rejected, we shall find time and again that his answers reflect the ethical norms of Christianity even when he has long since lost all touch with Christian images and parables. If the magnetic force that has guided this particular compass—and what else was its source but the central order?—should ever become extinct, terrible things may happen to mankind . . ." (p. 217[4]).
48. Inferred from the discovery of the consecutive formation of atoms in the expanding shells of quasars (Fig. II-2).

Glossary-Index[1]

THE NECESSITY FOR TWO GLOSSARIES.

The vocabulary of Unified Science is identical with those of the traditional sciences in regard to empirical data. Yet, while the properties and objectives which are peculiar to Unified Science are found in all of its empirical components, they result from the marriage of formal disciplines: of General Systems Theory with geometry. It is not, therefore, by chance that Unified Science appears, from the viewpoint of logic, to carry out well known proposals of two famous General Systems authorities, Ludwig von Bertalanffy and Kenneth Boulding.[2]

"General systems theory in the narrower sense (G.S.T.)," says von Bertalanffy, "is trying to derive from a general definition of 'system' a complex of interacting components, concepts characteristic of organized wholes . . . and to apply them to concrete phenomena."[3] This method is typical of the deductive-theoretical mode of thought.

A major objective of Unified Science, therefore, is to organize the verbal and visual symbols for this deductive operation. Its further objectives include the arrangement of the empirical data in such a way as to permit the attainment of what Kenneth Boulding regards as a major objective of Systems Theorists; namely, the transformation of the present aggregation of primarily empirical sciences into "a spectrum of theories—a system of systems."[4] This objective has, I believe, been reached in the present model of Unified Science by combining Boulding's two approaches in a concrete, practical way.

He sets them forth as follows: "The first is to look over the empirical universe and pick out certain general *phenomena* which are found in many different disciplines, . . . to build up general theoretical models relevant to these phenomena. The second approach is to arrange the empirical fields in a hierarchy of organization of their basic 'individual' or unit behavior, and . . . to develop a level of abstraction appropriate to each."

These two approaches result in the same thing, Unified Science: The Systems-hierarchy turns out to consist of Major Periods and Periodicity to recur within each Major Period, albeit in ever changing empirical forms. Boulding's two methods are, however, entirely diverse: the first belongs to the inductive-empirical, the second to the deductive-theoretical mode of thought, graphically shown in Figure 5-1.[5]

These two modes of thought call for diverse kinds of glossary: The inductive-empirical mode calls for the traditional alphabetically ordered kind of glossary; the deductive-theoretical mode calls for a holistic, graphic kind of glossary which permits the various relations of the necessarily complex theoretical construct, Unified Science, to be seen as they must be, simultaneously, each one then being defined in terms of others, as a mathematical equation is first presented in complete form, its components being defined sequentially thereafter.

Accordingly, Unified Science's glossary is presented in two complementary forms: The relational, graphic form for holistic coherence, and the alphabetical form combined with the index, for retrieval and the definition of terms not spelled out in the first form.

GRAPHIC, DEDUCTIVE-THEORETICAL GLOSSARY.[6]

(Theoretical Summary.)

A graphic representation of Unified Science results from the execution of Bertalanffy's and Boulding's above proposals, Figures 2-1a and 2-1b. Figures representing six major concepts are listed there sequentially. The main components of each one are then defined verbally as follows:

1) *General System* is defined verbally and visually, Figure 2-1a. Some of its components are defined in logical sequence.

2) *Empirical System:* any component of the left-hand column, Figure 2-1b, defined verbally in the alphabetical glossary.

3) *System-Hierarchy*, abstract (execution of Boulding's "system of systems,") left-hand column of Figure 2-1b, verbally defined here and in the alphabetical glossary.

Glossary — Index 217

4) *General Periodic Table* and the Taxonomic System of Unified Science. Figure 2-13 by Harold G. Cassidy. Verbal definitions of the whole and some of its components follow.

5) *Empirical Periodic Table* (paradigmatic form), item 3 in the center column of Figure 2-1b, shown in Figure 2-5a.

6) *Hierarchy of Empirical Periodic Tables* (execution of Boulding's "spectrum of theories" and "levels of theoretical discourse"), right-hand column, Figure 2-1b, and Unified Science chart.

DEFINITIONS OF THE MAJOR CONCEPTS' COMPONENTS

1) General (Theoretical) System (Figure 2-1a)

A space-time region bounded by sharp but not complete breaks of interdependence between its components and their environment, these incompletenesses being inputs and outputs. (E. F. Haskell and H. G. Cassidy, 1963).

a) *Interdependence* is a relation between two or more entities, where each causation or action produces one or more reactions, retro-actions, or feedbacks.

b) *Entity:* "A thing which has reality and distinctness of being either in fact or for thought; as, to view the state as an entity." *Webster's New Collegiate Dictionary*, Merriam Co., Springfield, Mass., 1957.

c) *Habitat:* All things affecting an entity and which it affects at the time in question. E. F. Haskell, *Ecology* 21, 1 (January 1940): "Mathematical Systematization of 'Environment,' 'Organism,' and 'Habitat.'"

d) *Environment:* All things which affect an entity's habitat or are affected by it. (*Ibid.*, and Figure 2-1a.)

e) *Output:* Energy, material, or information emanating from an entity or system during a given time span.

f) *Input:* Energy, material, or information entering an entity or system during a given time span.

2) Empirical System

A General System in which the abstract Entity is replaced by an empirical entity, and in which all other abstract components are replaced by empirical components implied by the definitions of *habitat* and *environment*. Examples: Any one of the empirical systems represented in the left-hand column of Figure 2-1b.

3) Systems-Hierarchy

"A hierarchy of empirical systems such that each member of the hierarchy (except the first) consists of some or all previous members of the hierarchy, plus one or more entities emerged from the hierarchy, mutually modified." (Formulated jointly by W. V. Quine, H. G. Cassidy, and E. F. Haskell, 1964.)[4]

The General Periodic Table (Figures 1–3 and 5–5)

A Systems-Hierarchy such that its emerging entity is its controlling component, that this component's emergence occurs in a regular manner, and that the system's over-all organization is correlated with that of its emergent entity, and varies as it does.[7] This is a paraphrase of the General Periodic Law, $R = f(\theta)$.

a) *Groups:* Classes of organizational relations between the activity of the system's work component X and that of its controller Y, Figures 1-2 and 1-3. There are nine Groups, consisting of the permutations and combinations of +, 0, and − for work component X and controller Y.[8] Geometrically, the Groups are classes of directions of the radius vector.

b) *Period.* (See Figure 2-13.) A numbered category of all natural systems classifications, whose number is equal to the number of Strata (q.v.) comprising the system in question.

c) *Stratum.* (See Figure 2-13.) A numbered category of all natural systems classifications. In atoms, for instance, the electron shells; in human cultures, the social strata.

Glossary — Index

d) *Sub-Stratum:* The developmental or ontogenetic stage of any individual member of a Stratum. For any given Stratum, the theoretically possible number of Sub-Strata is equal to the Stratum's own number (Figure 2-13.) This holds true for all Periodic Tables shown in the fold-out chart at the rear.

e) *The Taxonomic System* of unified science presupposes and includes the taxonomies of all the empirical sciences some of which, however, classify just system *parts*. It combines and orders these systems-theoretically and geometrically by means of Characteristic Numbers, as proposed by Leibniz in 1677.[9] Sub-Stratum, Stratum, Period, and Kingdom (or Major Stratum) jointly define a System's scalar location R, its outward position on the radius vector (see below); while Group defines the vector's direction. Any System's position in the Periodic coordinate system (Figure 5-5) may thus be rendered conveniently as follows: E, $\genfrac{}{}{0pt}{}{\genfrac{}{}{0pt}{}{S}{S\text{-}S}}{P}$ K G

where E is the entity's traditional taxonomic designation.

    K is its Kingdom or Major Stratum within the Systems-hierarchy

    P is its Period within its Major Stratum

    S is its Stratum within its Period

  S-S is its Sub-STRATUM WITHIN ITS STRATUM

    G is its Group within the Periodic coordinate system

The first four geometric parameters (K through S-S) define the entity's scalar position R_0 and are written with Arabic numerals.— The fifth parameter, G (Group), defines its radial position θ, and is written with a Roman numeral, following the tradition established by D. I. Mendeleev.—When it is more convenient to write characteristic numbers sequentially, rather than in the pattern above, that order may be preferred.

5) The Paradigmatic Empirical Periodic Table

Figure II-5a and b, the Periodic Table of Chemical Elements. In this Systems-Hierarchy the controlling component, whose emergence is strategic, is the nucleus. Its emergence occurs in expanding quasar shells by sequential additions of one proton. And in this System's (the atom's) over-all organization, the structures of both its nucleus and of its work component the electron cloud, are correlated with the quantized emergence of the nucleus, and vary as it does. (Paraphrasis of Mendeleev's Periodic Law.)

- P) *Period:* The class of atoms having any given number of Strata (nuclear and electron-cloud shells) in common.
- S) *Stratum:* A nuclear shell, an electron shell, and usually both together.
- S-S) *Sub-Stratum:* A set of states which may be assumed by entities comprising any given Stratum.
- G) *Groups:* Classes of organizational relations between the atom (entity) and its habitat. (Sets of chemical properties.)

Notation: All characteristic numbers in which $K = 2$ (the second Major Stratum).

6) Hierarchy of Empirical Periodic Tables

Right-hand column in fold-out Unified Science chart, (fold-out chart situated at the end); also in Figure 5-5.

Just as the graphic representation of the classes of empirical systems forms a Systems-Hierarchy ("system-of-systems"), so that of their Periodic classifications forms a Systems-Hierarchy of Periodic Tables ("spectrum of theories"). This hierarchy will eventually display seven; perhaps even eight members if patrons are verified. They constitute the Systems-Theoretic counterpart of what are traditionally called *natural kingdoms* and we call Major Strata. The symbol proposed for them is therefore K.

K 1) Periodic Table of Stable Particles. (It has 1 Stratum, 1 Period: Alpha coordinate system in fold-out chart.)

K 2) Periodic Table of Chemical Elements: Periodic coordinate system, Table of Atoms.

K 3) Predicted Periodic Table of molecules.

K 4) Predicted Periodic Table of geoid systems.

K 5) Periodic Table of Biopetic Ecosystems (Phylogeny of the cell by H. G. Cassidy) extends into the Periodic Tables of Plant Ecosystems.

K 6) Periodic Table of Animal Ecosystems.

K 7) Periodic Table of Human Cultures.

Glossary — Index 221

7) **Role of Traditional Taxonomies** (Figure 2-1b, center column).

This Deductive-Theoretical Glossary sums up the present execution of Bertalanffy's program for General Systems Theory.[1] This result was obtained by combining the two methods which Boulding proposed, and apparently results in a model of his system-of-systems and spectrum-of-theories.[2]

Accordingly, the Periodic coordinate system's coordinates, scalar and polar, represent levels of organization, whose lower limit *Alpha* is the point of maximum disorganization and whose upper limit Omega is the region of maximum organization, Figure 5-5.

Between these limits are ranged the entities represented in all three columns of Figure 2-1b: material entities in the left-hand column, unassembled theoretical sub-assemblies in the center, and assembled sub-assemblies in the right-hand column. Their build-ups and breakdowns can be represented, diagnosed, predicted, prescribed; and executed in terms of the Periodic coordinate system, and the results compared with the theoretical realities in question. This provides the general background theory and meta-language which permit the reform of the crisis-ridden Multiversity into the New University, Figures 4-9 and 4-11. For the boundaries between the natural kingdoms, between the disciplines describing them, and the information and organization distributed among these boundaries are hereby assembled into the panmathic discipline, Unified Science, obtained by executing von Bertalanffy's and Boulding's proposals.

NOTES AND REFERENCES
ALPHABETICAL GLOSSARY AND INDEX

1. Adapted from the *Glossary Proposal for General Systems Theory*. This proposal was originally drawn up by a vocabulary committee consisting of George J. Klir and Edward Haskell. This committee was set up at New Haven in November 1968 by the Task Force on General Systems Education of the Society for General Systems Research, chaired by Jere W. Clark.—This paper with the kind consent of Dr. Klir and Dr. Cassidy.
2. Historically, this model executes logically similar, but less well formulated proposals: it was completed before the present proposals' publication in 1968.
3. Ludwig von Bertalanffy, "General Systems Theory—A Critical Review" in *Modern Systems Research for the Behavioral Scientist*, Walter Buckley, Ed., Aldine, Chicago, 1968.

4. Kenneth E. Boulding, "General Systems Theory—The Skeleton of Science," in *Modern Systems Research for the Behavioral Scientist*.
5. James B. Conant, *Two Modes of Thought—My Encounters with Science and Education*, R. N. Anshen, ed., Trident Press, New York, 1964.
6. This appears to be a novel kind of glossary. If it needs justification, this may be found in *Imagery and Verbal Processes* by Allan Paivio, Rinehart and Winston, New York, 1971.
7. This concept was formulated by John Stuart Mill in 1868 in his *Philosophy, Ratiocinative and Inductive*, under the term Natural Classification. The following year Mendeleev and Maier discovered the first empirical Periodic Table independently of him and of each other. Four more natural classifications were discovered a little less than a century later, and two missing ones predicted, Unified Science Chart.
8. In empirical cases Periods do not necessarily consist of 9 Groups: There are Periods in which some Groups are omitted (the first in the Chemical Table has 2 Groups); some in which Groups are duplicated (e.g., the "long" atomic Periods). The General Periodic Table, however, (Figure I-3), is basic to the understanding of General Groups and Periods.
9. Leibniz Selections, Philip P. Wiener, editor, Scribners, New York, 1951 (pp. 12–25).

Glossary

Abiotic (adj.) Non living. The class of systems which includes particles, atoms, molecules, and geoid (q.v.) systems. C.f. Biotic, cultural.

Abstraction ceiling The highest abstraction-level (q.v.). In ontogeny, the abstraction ceiling is the individual's highest level at the time in question, and is indicated at the left of his Characteristic Number (q.v.). The Stratum ceiling is the highest level reached by the average Stratum member, and is indicated at the top of his characteristic Number. The Period ceiling is the highest level reached by the Period's top Stratum, and is indicated at the bottom of the characteristic number. (*See* Figure IV-2.)

Abstraction, levels of In sentient systems (*res cogitans*), which conduct simulation of systems behavior, the controller reflects upon, *abstracts* simulated outputs (*see* Feedback, double, triple, etc.), then abstracts simulation of its reflection, and so forth in an abstraction hierarchy. This is a Systems-hierarchy (q.v.) composed of levels of abstraction. (*See* Feedback, double, triple, etc.)

Allopathy (n.) (Greek for "making the other sick"; coined term) The coaction (q.v.) in which the activity of the system's work component (q.v.) is unaffected, but that of its controller (q.v.) is decreased. *Allopathy*, written (0, −), is the characteristic of Group I in all Periodic Tables. (It was first discovered theoretically, then empirically.)

Allotrophy (n.) (Greek for "feeding the other; coined term) The coaction (q.v.) in which the activity of the system's work component (q.v.) is not affected while that of the controller is intensified. *Allotrophy* is written (0, +). It is the characteristic of Group V in all Periodic Tables. (It was first discovered theoretically, then found empirically in animal and human ecosystems, q.v.).

Alpha, A The point of minimum organization or maximum disorganization. The symbol situated at the center or origin of the Periodic coordinate system (q.v.); its lower limit. C.f. Omega

Alpha coordinate system (coined term) The coordinate system which forms the transition between the Periodic and the Inverted Periodic coordinate systems (q.v. and Figure II-10). The Cartesian coordinate system with coordinate axes reversed, into which the Periodic table of stable particles (q.v.) can be mapped.

Amensalism (n.) (coined term) The coaction (q.v.) in which the activity of the work component (q.v.) is decreased while that of the controller is not affected. Amensalism, written (−, 0), is the characteristic of Group VII in all Periodic Tables. (It was first discovered theoretically, then empirically.)

Atropy (n.) (coined term) Stability; neither increasing nor decreasing organization (q.v.) C.f. Entropy, Ectropy.

Atropy, circle of *See* Reference zero.

Background theory (term coined by W. V. Quine 1969) An abstract theory from which diverse, less abstract theories can be inferred. Conversely, a theory implicit in diverse less abstract theories.

Being, extensive (*Res extensa*) Having extension, number, speed, velocity, mass, momentum. In general, the class of abiotic and low biotic systems (q.v.). C.f. Being, sentient, *Res cogitans.*

Being (n.) sentient. (*Res cogitans*) Sentience is the capacity of a being or system to simulate systems behavior, as also to simulate simulation; that is, to simulate the behavior of sentient systems, living or mechanical, such as computers. Sentient systems thus form a systems-hierarchy (q.v.) such, that each higher member of the hierarchy accurately simulates the behavior of lower members than itself. (*See* Simulation.) The hierarchy evolves by emergence (through mutation, maturation, training or invention) of ever higher levels of abstraction (q.v.). Simulation of the behavior of higher by lower members of the hierarchy is called mimesis (q.v.). The part of the universe excluded from Western science in the 17th century by mutual consent of scientists and theologians.

Biocenose (n.) A living community, usually including abiotic, plant and animal components. (*See* Biome, Ecosystem.)

Biome (n.) A major community of living organisms; a complex of mature communities of plants and animals in a major region, as tundra, forest, grassland.

Biotic (adj.) Living (viral, plant, animal, or human). C.f. Abiotic, cultural.

Breakdown Sudden decrease of organization (q.v.). C.f. Disintegration.

Characteristic, Universal The characteristics which all systems (q.v.), especially all natural systems, have in common. Term coined by Leibniz ca. 1677, who predicted that the universal characteristic would be formulated geometrically. This prediction appears to have been fulfilled by the Periodic coordinate system (q.v.) which purports to be a model (q.v.) of the universal characteristic.

Characteristic number (*See* Number, characteristic).

Circle of atropy (coined term) Any one of the hierarchy of concentric circles centered at Alpha, the origin of the Periodic coordinate System (q.v.). Geometrically, a circle relative to which entropy and ectropy (q.v.) can be defined and represented. The region in which the (0, 0) Group (q.v.) of every Period is mapped. Hence, it is also called the *zero-zero circle* or the *scalar zero circle.*

Class conflict (n.) Synnecrosis, predation or parasitism (q.v.) between a civilization's Majority or work component (q.v.) and its Minority or controller (q.v.). A set of coactions typical of a culture's breakdown and disintegration (q.v.); of its Alpha-ward devolution (q.v.). The condition typical of Dionysian cultures; e.g. the Kwakiutl Indians or current Northern Ireland. In Marxist theory, "All of recorded history is the history of class conflict". This misinterpretation of history results from this theory's negative value-bias (q.v.). C.f. Front, horizontal.

Class cooperation (n.) Symbiosis (q.v.) between a civilization's Majority or work component (q.v.) and its Minority or controller (q.v.). This coaction (q.v.) is essential to a culture's genesis, stability and upward (Omegaward, q.v.) development. The normal condition of Apollonian cultures; e.g. the Pueblo Indians or modern Switzerland. (*See* Figure V-4). A condition termed "impossible" by Marxist theory, because the negative value-bias (q.v.) of its formal structure makes it (subjectively) "impossible". C.f. Front, vertical; "Classless" society; *contrast with* Class conflict; Front, horizontal.

"Classless" society (n.) Marxist concept of a "future state of society" in which Majority or work component (q.v.) and Minority or controller (q.v.) are affirmed to "disappear". This concept, which contradicts systems theory and cybernetics and has no empirical support, was substituted for the concept of class cooperation (q.v.) because the negative value-bias (q.v.) of Marxist theory precludes the concept of class cooperation or symbiosis (q.v.). "Classless" society therefore is always said to exist in the "future". Existing socialist societies are necessarily governed by a controlling class. (*See* Milovan Djilas' *New Class*.) But since the possibility of class cooperation is precluded by the incorrect structure of Marxist theory (*see* Chapter V), this controller is given another name: "Vanguard of the Proletariat". Workers who revolt against this controller, called Leftists, are especially strong in France.

Coaction (n.) Any of the nine theoretically possible types of relation between a system's work component and controller in regard to any given activity or goal (Tel). Coaction is defined as the direction of any given system's radius vector. C.f. Coaction compass; Value, moral.

Coaction Cardioid The generally heart-shaped path of the radius vector in the Periodic Coordinate System, whose equation is $R=f(\theta)$. (C.f. Periodic Law, General). The coaction cardioid turns into the zero-zero or scalar zero circle (q.v.) in the region of predominantly negative coactions (*inturning*, in Greek, is *entropy*) toward Alpha; and turns out of the circle in the region of predominantly positive coactions (*turning out*, in Greek, is *ectropy*, q.v.) toward Omega (q.v.). The coaction cardioid thus crosses the circle of atropy (q.v.) twice.

Coaction Compass A geometric coordinate system which defines the nine possible coactions and shows the relations between them. C.f. Periodic Coordinate System.

Commensalism (n.) (Latin for togetherness at the table) The coaction (q.v.) in which the activity or goal-achievement of the system's work component (q.v.) is intensified while that of its controller is not affected. Commensalism, written $(+, 0)$, is the characteristic of Group III of all Periodic Tables.

Complexity The property of having intricate structure and operation, predisposing a system to breakdown (q.v.). A condition of a system requiring continuous organization (q.v.) in the form of servicing and maintenance. Hence often confused with organization, especially by social scientists.

Control (n.) The converse of function in the systemic and mathematical sense (q.v.). E.g. If $R=f(\theta)$ then θ has or is the control of R: $\theta=c(R)$. Control is to "Dominate, command; hold in check (oneself, one's anger); check, verify". *Concise Oxford Dictionary*, 1942. Human control may occur through wisdom, persuasion, deception, conviction, the ballot, or physical force. It is displayed by the Minority (q.v.), whether Capitalist, Communist, fascist, Social Capitalist (q.v.), or other. C.f. Controller; Moral force.

Controller (n.) The component in a cybernetic system which reacts to a change in the output, providing a signal that alters the output toward the system's established goal. The more strategic entity in an ordinated relationship (q.v.) between the two basic components of cybernetic systems, the less strategic being its work component. C.f. Control, Ordination, Work component, Strategicity.

Coordinate system Coordinates are numbered points on a straight line or axis. A coordinate system can be one-dimensional, e.g. the (spurious) political "spectrum"; two-dimensional, e.g. rectangularly crossed X and Y axes; three-dimensional, e.g. rectangularly crossed X, Y and Z axes; and so on to n-dimensional. C.f. Periodic coordinate system.

Cultural (adj.) (human). C.f. Abiotic, Biotic.

Cybernetics (n.) The science of behavior, communication, control and organization (q.v.) in organisms, machines, societies and other systems. Its salient characteristic is feedback or retroaction (q.v.).

Devolution (n.) Degenerative development, gradual or sudden, toward a lower level of organization; geometrically, in the direction of Alpha (in the Periodic coordinate system). C.f. Breakdown, Disintegration.

Disintegration (n.) The transmutation of a system, such that it ceases to exist and appears as disintegration products; that is to say, into systems belonging to one or more Periods or Major Periods (q.v.) lower in the systems-hierarchy (q.v.).

Ecology (n.) The science dealing with ecosystems: Entity-habitat systems with imputs from and outputs to their environments (q.v.).

Ecogeny (n.) Formative change or development of ecosystems.

Ectropy (n.) (Term coined by W. V. Quine, 1969.) Increasing organization. In Greek, *ectropy* means *turning out*. Increasing ectropy is represented in Periodic geometry (q.v.) by the coaction cardioid's (q.v.) *turning out* of the reference-zero circle (q.v.). (*See also* Entropy, Atropy.)

Effector, Constancy A cybernetic system which holds close to a norm or constant goal. C.f. Tendency Effector.
Pierre de Latil *Thinking by Machine—A Study of Cybernetics*. Saunders, London, 1968.

Effector, Tendency A cybernetic system which tends away from its norm toward a maximum or a minimum, e.g. an atomic bomb. C.f. Effector, constancy.
Pierre de Latil *Thinking by Machine—A Study of Cybernetics*. Saunders, London, 1968.

Glossary

Empire, natural (coined term) A member of the system-hierarchy of empirical systems (Fig. II-1b) consisting of ("containing") all lower members of the hierarchy plus an emerged controller. Each natural Empire or Major Period (q.v.) bears the name of its highest major Stratum (q.v.) or natural kingdom (q.v.). Thus, the natural Empire of man is his ecosystem, the solar system.

Entropy (n.) Increasing disorganisation. In Greek, *turning in*. See Co-action cardioid; Ectropy, Atropy.

Environment (n.) In Unified Science, all things which effect or are effected by the component of a system called *habitat* (q.v.). C.f. Figure II-1a. Thus, *environment* bears the same relation to *habitat* which *habitat* bears to entity, generating a regressive series of ecosystems (q.v.) whose empirical limit is an ecosystem with unorganizedp articles as *entity*, Alpha as *habitat* (q.v.).

Evolution (n.) The change, gradual or sudden, of any entity's organization (q.v.) in the direction of higher organization. See Ectropy, Omega. See also Devolution.

Evolution, abiotic Evolution (q.v.) of non-living systems such as atoms, molecules or geoid systems. See fold-out chart and Figure II-2.

Evolution, Bio- Evolution (q.v.) of biotic or livng systems.

Exploitation (n.) A popular but operationally inconstant term used to denote both predation (q.v.) and parasitism (q.v.). Sometimes even extended to X-aggerated symbiosis $(+ > +)$ and Y-aggerated symbiosis $(+ < +)$.

Feedback (n.) The returning of a portion of the output of a system to the input for purposes of control. Negative feedback directs the output closer to the system's goal or norm. (*See* Constancy effector.) Positive feedback directs the output away from the goal, toward infinity or zero. (*See* Tendency effector). Synonym: retroaction. These definitions are by Pierre de Latil (*Thinking by Machine*).

Feedback, double, triple, etc. A closed system of simulated responses which enables a system to reflect upon perceived phenomena and also upon the simulated actions called thoughts. In this manner, a number of possible actions can be imagined and the best one implemented: the one which will presumably come closest to the system's goal. Feedbacks may be double, triple, quadruple, etc., according to the number of levels of abstraction involved. C.f. Abstraction, levels of.

Feedback, negative Feedback which decreases the deviation of a system's output from its norm or goal. C.f. Constancy effector.

Feedback, positive Feedback which amplifies the deviation of a system's output from its norm or goal. C.f. Tendency effector.

Feed-forth (n.) (term coined by W. V. Quine.) A signal generated by a system which changes its norm or goal (usually by changing the habitat). E.g. in *biotic succession* (q.v.) each stage of the System-hierarchy (q.v.) prepares the habitat by *feed-forth* for the next stage.

Fouling (of information) (n.) Operational inconstancy of signals which diminishes, or destroys the information they convey. C.f. Signal, Noise.

Framework (n.) (of a Periodic table) The formal structure which all Periodic Tables have in common.

Front, horizontal (coined term) A split in a society between work component (q.v.) or lower Strata (q.v.) and controller (q.v.) or upper Strata (q.v.). The front formed by class conflict (q.v.). In anthropology, *horizontal schizmogenesis* (Gregory Bateson), which results in social disintegration. The method by which Marxists seek to alienate a society's work component (q.v.) from its controller (q.v.), and then to take its place. *Compare with* Front, vertical.

Front, vertical (coined term) A split in a society from top to bottom such, that a part of the work component (q.v.) and of the controller (q.v.) remains intact on each side of the vertical front. In anthropology, *vertical schizmogenesis* (Gregory Bateson), which results in regeneration of society. The method by which Social Capitalists (q.v.) transform exploiters (predators or parasites, q.v.) into symbionts or class cooperators (q.v.), thereby preventing class conflict (q.v.) and building stable and healthy societies and ecosystems. E.g. modern Switzerland. C.f. Social capital. *Compare with* Front, horizontal.

Function (n.) "Activity proper to anything, mode of action by which it fulfils its purpose . . . (math.) variable quantity in relation to other(s) in terms of which it may be expressed or on which its value depends." *Concise Oxford Dictionary*, 1942. The converse of control (q.v.).

Galaxy (n.) The category whose members are the largest known material components of the universe. The highest known stage in the development of a quasar (q.v.). (*See* Figure II-1b.)

Genesis of civilization (Ref. Toynbee, A. J.) The state of a human society in which the prevalent *subjective coaction* (q.v.) between Majority or work component (q.v.) and Minority or controller (q.v.) is *positive* (q.v.): the state called *trust, confidence,* class cooperation (q.v.). During *genesis*, a society is characterized by *stability* and *upward development* (q.v.). Genesis of society is always characterized by a *public philosophy* (q.v.) with a positive *value bias* (q.v.).

Geoid (adj.) (Term coined by W. V. Quine, 1969) The class of systems (the Major Stratum q.v.) occupying the position in the System hierarchy (q.v.) between molecules and plant ecosystems. Among its members are inter-stellar gas clouds, stars, meteors and solar systems. A Periodic table of geoid systems is predicted.

Group (with capital G) (n.) Originally, any one of the columns of the Periodic table of chemical elements, (q.v.), all of whose entries have chemical properties in common. Here extended to, all Periodic tables, abiotic, biotic, and cultural (q.v.), and the Periodic coordinate system into which they map sequentially. (*See* fold-out chart).

Habitat (n.) (Definition by Haskell, 1940) In Unified Science, all things which affect or are affected by an entity (q.v.) comprise its habitat.—*Habitat* and *entity* together comprise the *ecosystem* (q.v.). C.f. Environment.

Homeostasis (n.) Maintenance of conditions in a constant state; e.g. the temperature or humidity of a system. C.f. Constancy effector, Negative feed-back.

Glossary 229

Information (n.) Structure or organization transmitted by means of a signal (q.v.).

Kind, natural (term coined by W.V. Quine, 1969) A class of natural systems.

Kingdom, natural A large category of traditional science. In Unified Science, the Major Strata (q.v.): 1 particles, 2 atoms, 3 molecules, 4 geoid systems, 5 plant ecosystems, 6 animal ecosystems, 7 human cultures. In the *Characteristic Number* (q.v.), the central number. C.f. Stratum, major; *compare with* Period, major; Kind, natural.

Law, General Periodic $R = f(\theta)$ where R is the length of the radius vector and represents the system's properties, and θ is the radius vector's direction and represents the system's coaction or value-bias (q.v.). Also called the Moral Law.

Mimesis (n.) (Ref. Toynbee, A. J., *A Study of History*) Simulation, within a Systems-hierarchy (q.v.), of the behavior of higher members by lower members of the hierarchy. Since lower members by definition lack the structure which enacts the system's highest levels of abstraction (q.v.), mimesis is confined to those aspects of behavior which lie beneath their abstraction ceiling (q.v.). *Mimesis* is seen by higher members of the hierarchy as inept, valuable, dangerous, ludicrous, cute, etc. depending upon the situation. None the less, mimesis is essential to the coherence and operation of societies in all human Periods: Even in the lowest, it characterizes much activity of children and young people (lower Sub-strata).

—Understanding of mimesis by the top Strata is important to society's control, confidence and trust (q.v.), which are necessary conditions for social stability and Omega-ward development. C.f. Class cooperation, Symbiosis, vertical Front.

Minority, creative A controlling class in a society which cooperates with the society's working class, thereby either transmuting the whole society to a higher Period (q.v.) or forming a *vertical front* (q.v.) and transforming predation and parasitism (q.v.) into symbiosis (q.v.). Term coined by Arnold Toynbee. C.f. Minority, dominant, Control, function.

Minority, dominant A controlling class in a society which exploits the society's working class, preparing conditions for the formation of a horizontal front (q.v.) and disintegration (q.v.) of society. Term coined by Arnold Toynbee. C.f. Minority, creative, Control.

Model (n.) The result of the *simulation* (q.v.) of an idea or object. *Example:* The Periodic coordinate system is a *model* of Leibniz's Universal Characteristic (q.v.). Also, and conversely, the object of simulation, as in the *prototype model* of a car or aeroplane.

Moral force The capacity to direct or orient a system's coactions (q.v.) or quality (q.v.), thereby to control its properties (q.v.) C.f. Moral Law, Periodic Law, General.

Moral Law The general law of which Mendelee'v Periodic Law is the special atomic case. Traditionally, "As ye plant, so shall ye reap." Cybernetically, "The properties of systems are functions of coaction," (q.v.). C.f. Moral force, Quality, Quantity.

Noise (n.) As used in information theory: impurities in a signal (q.v.), including fouling (q.v.), which diminish or eliminate its information (q.v.).

Number, characteristic Originally proposed by Leibniz (1677). The elementary unit of the taxonomic system of Unified Science. Any entity may be classified, and its data retrieved, by means of its Characteristic number. This consists of the five parameters which map it into the Periodic coordinate system (q.v.). These parameters are: K (the entity's natural kingdom or Major Stratum, and thus its Periodic table), P (its Period in its own Periodic table), S (its top Stratum or ontogenetic ceiling), S-S (its sub-Stratum or ontogenetic stage at the time in question), and G (its Group or characteristic coaction at that time). (*See* these glossary entries).

Number system, Natural The number system of the Periodic coordinate system (q.v.). The oldest number system, which originally began with 1, and extended to, positive infinity. Today it begins with zero. C.f. Number system, Real.

Number system, Real The number system of the cartesian and the Alpha coordinate systems (q.v.). The number system brought by Arabs from India to the West. It includes zero and both positive and negative numbers. C.f. Number system, natural.

Omega, Ω The outer limit of the Periodic coordinate system (q.v.). The point of maximum organization (q.v.). Originally proposed by Teilhard de Chardin as the state toward which evolution is directed. Compare with Alpha.

Ontogeny (n.) Development of the individual organism—plant, animal, or human—from zygote (q.v.) to death. C.f. Ontogeny, stage of, ceiling of.

Ontogeny, ceiling of The highest stage in the individual's or entity's ontogeny or life-development—whether human, animal, plant, geoid, molecule, atom or particle. This ceiling determines the individual's Stratum (q.v.).

Ontogeny, stage of Development stage of the individual or entity (particle, atom, molecule, geoid system, plant, animal, or person). In all cases, the individual's ontogeny recapitulates its species' evolution or phylogeny (q.v.). This is a concrete formulation of Haeckel's Law, "Ontogeny recapitulates phylogeny". C.f. Substratum and Ontogeny, ceiling of. Note, however, that specifically *human* ontogenetic stages, as here expounded and numbered, are mental. (C.f. Abstraction, levels of.) Preceding stages, those of embryonic development, are shared with the highest animals.

Ordination (coined term as here used) (n.) The relation of greater, equal, or less, whose mathematical symbol is the combination of $<$ or $>$ and $=$; namely $< = >$. *Ordination* is particularly relevant to the relation of work component and controller in a system, because of its immediate relation to the General Periodic Law (q.v.).—If coaction is positive and *ordination* is upward, (such that lower Strata and Sub-strata are acting as work component X, higher as controller Y ($X < Y$)), the system is in the state of *evolution* or *genesis* (q.v.). If *ordination* is equal or downward ($X \geqslant Y$), the system is in the state of *devolution*; of

Glossary

breakdown or *disintegration* (q.v.).—This is a paraphrasis of the Second Periodic Law*, of which the Second Law of Thermodynamics is the special case $(X > Y)$.—Systems which appear to have no work component or controller may thus be viewed as members of a second special case $(X = Y)$. The third case $(X < Y)$ involves increasing organization or ectropy (q.v.).

Organization This property of systems is so fundamental as to be logically undefinable. It is the condition whose satisfaction by any set of system-components results in the next higher member of the System-hierarchy (q.v.); and whose non-satisfaction breaks it down into component systems lower in the hierarchy. Hence its theoretical maximum and minimum—called Alpha and Omega (q.v.)—constitute the limits of the Periodic coordinate system (q.v.), the framework of Unified Science.—Often confused with complexity, q.v.

Parasitism (n.) As used in Unified Science, the coaction in which the activity of the system's work component (q.v.) is increased while that of its controller (q.v.) is decreased. Parasitism, written $(+, -)$, is the characteristic of Group II in all Periodic Tables. Noisy Synonym (c.f. Noise): exploitation.

Period (with capital P) (n.) (1) The scientific version of recurrences noted by poets and philosophers both in nature and in human history. (2) In the seven Periodic Tables (q.v.), the rows composed

---

*$R = f(< = >)$, where R is the property called *volution*, the class whose members are *e*volution, *de*volution, stability.

of the Groups (q.v.) (D. I. Mendeleev, 1869.) (3) In the Periodic coordinate system (q.v.), the concentric reference-zero circles (q.v.) and their accompanying coaction cardioids (q.v.). C.f. Period, major.

Period, major (coined term) An entire Periodic Table, which constitutes a Major Period of the Major Periodic Table (q.v.). See Empire, natural; compare with Kingdom, natural and Stratum, Major.

Periodic coordinate system The coordinate system of Unified Science. (*See* Chapters I and II). Obtained by generalization of the Periodic Table of chemical elements. The coordinate system into which all Periodic Tables (q.v.) can be mapped hierarchically.

Periodic geometry The geometry related to the coordinate systems of unified science and represented by the Periodic, the Alpha and the Inverted Periodic coordinate systems (q.v.). Generated by combining cybernetic with geometric principles.

Periodic Law, chemical The properties of the chemical elements are Periodic functions of their atomic numbers. (D. I. Mendeleev, 1869). A special case of the General Periodic Law (q.v.).

Periodic Law, General $R = f(\theta)$, where $R =$ length of the radius vector and $\theta =$ radius angle. Stated in words: the properties of systems are functions of their coactions (q.v.). Synonym: The Moral Law of Unified Science. *See* Periodic Law, chemical.

Periodic Table, empirical A Table or framework (q.v.), into which has been mapped a natural kind (q.v.) of type specimens (q.v.). E.g. The Periodic Table of chemical elements, of animal ecosystems, or of human cultures. C.f. Periodic Table, General.

Periodic Table, General (capitalized) A table whose columns are Groups (q.v.) and whose rows are Periods (q.v.), and which may include Strata and Sub-Strata (q.v.). Sometimes called the *framework* of the Periodic Table (q.v.). C.f. Numbers, Characteristic; also, Periodic Table, empirical, Major

Periodic Table, Major The Periodic Table of unified Science whose Periods are called Major Periods (q.v.) and are represented by Periodic Tables (q.v.).

Periodic table of stable particles (coined term) The Periodic table whose Groups contain the four stable pro-particles, four stable anti-particles, and the photon which is its own antiparticle. C.f. Alpha coordinate system.

Phylogeny (n.) Development of a species—plant, animal, or human —from emergence to extinction C f ontogeny.

Predation (n.) The coaction (q.v.) in which the activity of the system's work component is decreased while that of its controller is increased. Predation, written (−, +), is the characteristic of Group VI in all Periodic Tables. Noisy synonym (C.f. noise): exploitation.

Properties, systemic   A system's pro-properties are directly ascertainable and vary from one Major Stratum to another (q.v.). E.g. the properties of the chemical elements, of animal ecosystems, of human cultures. A system's properties are functions of the system's quality (q.v.). C.f. Periodic Law, chemical, General; Moral Law.

Public philosophy (Ref. Lippmann, Walter, *The Public Philosophy*.) A society's commonly shared world view. The higher the society's Period, the more forms its public philosophy must display: A different form for each Stratum. Coexistence of these different forms of the public philosophy depends upon the emergence of coherent Systems theory, complete with glossary, permitting translations between the diverse forms of the public philosophy. Unified science, religion and political ideology thus comprise the *public philosophy* of the Space Age. (Ref., Haskell, E. F.).

Quasar (quasi stellar object) (n.) The first known stage in the development (ontogeny, q.v.) of a galaxy. The population of atoms emerges in the quasar's expanding spheroidal shells. (See Figs. II-2 and II-1b) C.f. Galaxy.

Quality (n.) A system's quality is the principal coaction (q.v.) of its work component and controller (q.v.), and determines the system's properties (q.v.). Quality is geometrically represented as the radius vector's direction, appears in all Periodic tables as the Groups or columns, and is represented in Characteristic Numbers (q.v.) by the Roman numeral. C.f. Quantity, Periodic Law, chemical, General; Moral Law.

Quantity A system's quantity or) size) is equal to the cumulative sum of its Sub-strata, Strata (or Periods), and Major Strata (q.v.), represented by the four Arabic numerals in all Characteristic Numbers (q.v.). It is geometrically represented as the radius vector's length (q.v.), and appears

Glossary

in all Periodic tables as the rows or Periods. C.f. quality Periodic Law, General; Moral Law.

Recapitulation (n.) (As here used, coined by Ernst Haeckel ca. 1905.) Pertaining to the theory that individual life history (ontogeny, q.v.) enacts certain stages in the life history of the species (phylogeny, q.v.).
The individuals in question may be abiotic, plant, animal or human.

Radius vector (n.) A vector (q.v.) extending outward from the origin or center of a coordinate system. In the Periodic coordinate system (q.v.), the radius vector's direction defines the coaction or quality (q.v.) of the system in question—which in turn determine its properties (q.v.)—while its length defines the system's quantity or size (q.v.). C.f. Periodic Law, General.

Reference zero (term coined by H. G. Cassidy, 1972) The point or, in the Periodic coordinate system, the circle relative to which positive and negative qualities (q.v.) or coactions (q.v.) can be defined.

Retrogression (n.) Devolution to a lower, less complex type of individual or race; or type of behavior. C.f. Disintegration.

Scalar zero circle (coined term) *See* Circle of atropy, and Zero-zero circle.

Schizmogenesis (n.) (Term coined by Gregory Bateson.) Creation of a split in a system. *Horizontal* schizmogenesis splits the work component from the controller and results in mutilated components. *Vertical* schizmogenesis splits both controller and work component, resulting in two healthy daughter systems. (Gregory Bateson, *Naven*). C.f. Front, vertical, horizontal.

Science, Unified A new discipline, popularly known as *Unisci*. Obtained by mapping the data of the physical, biological and psycho-socio-political sciences into the Periodic coordinate system (q.v.). Proposed by Leibniz under the term *Scientia Generalis* 1677.

Selectee (n.) The entity selected. An organism selected for survival by its habitat (q.v.). Equally, a habitat selected by an organism. *See* Selector.

Selector (n.) The selecting entity. A habitat which selects an organism for survival. Equally, an organism which selects a habitat (q.v.).

Signal (n.) A physical carrier o information. A communications input into a system which changes its behavior or prevents change which would otherwise occur. Signals vary from level to level of the System-hierarchy (q.v.): Below the level of sentient beings (q.v.), power signals and communication signals are identical. Above the lowest level of sentient beings (which include higher animals) the ratio or signal input to control output increases in quantum *jumps* form level to level of abstraction. (*See* Abstraction, levels of).

Signalloid (adj.) (coined term.) An organism that is capable of receiving signals at a distance. Criterion of the second and higher Periods of animal ecosystems.

Signalzoa (n.) (coined term) Animals that use light, heat, chemicals, etc. as information to approach or avoid things, rather than as sources of energy only. C.f. Signaloid, Beings, sentient.

Simulation (n.) The matching of a system's structure or behavior by another system such, that some degree of isomorphy or structural correspondence of the two behaviors or structures results. (*See* Model, Modelling.) *Simulation* differs from *duplication* (q.v.) in that the medium in which, and method by which, the model is *simulated* is different from the materials of which the model is composed or enacted, and the method by which it was developed. In *duplication*, materials and methods are substantially the same in all resultant isomorphic systems. Simulation can be techno-*logical* (primarily controlled by mind) or techno-*genetic* (primarily controlled by the genetic apparatus). Some insects simulate plants technogenetically. Some computers simulate webs of life (q.v.) technologically. (*See* Technology, Technogeny.)

Social Capital (term coined by Gottlieb Duttweiler ca. 1935) A free-enterprise or capitalist economy displaying a vertical front (q.v.). Here, one side of the vertical front is traditionally capitalistic, displaying both predation and parasitism, (q.v.); the other side is symbiotic (q.v.) or class-cooperative. It is controlled by a creative Minority (q.v.) and competes with the part of society controlled by the dominant Minority (q.v.), forcing it into class cooperation $(-, +) \to (+, +)$ or $(+, -) \to (+, +)$. Social capital is a peaceful and sucessful alternative to socialism. C.f. Schizmogenesis, Front, vertical, Front, horizontal.

Strata (written with a capital S) The layers of levels in a system whose number determines the system's Period number. E.g. in atoms, the number of the atom's electron shells; in plant ecosystems, the number of the ecosystem's plant divisions. In each case, Period 3 has 3 Strata: in atoms, the K, L, M shells; in plant ecosystems, the algal, moss, and fern Strata. To be clearly distinguished from merely geographic strata (written with a small s), as in geologic strata or the strata of a forest. (*See* Sub-strata, Periods. *Also* Numbers, Characteristic.)

Strategicity (n.) (coined term) The quality of an entity within a system, of being so related to the other entities in it that a relatively small amount of energy produces a relatively great effect. Control is thus proportional to strategicity. C.f. Controller, Work component, Moral force.

Stratum, Major Synonymous with natural kingdom (q.v.) and represented by a Periodic Table (q.v.). C.f. Kingdom, natural.

Stratum, potential The individual's highest inborn ontogenetic stage (q.v.) or ceiling (q.v.); the genetic capacity whose realization, in living things, requires suitable habitat (q.v.) conditions to become actual or kinetic. In azoic systems, of course, the same. C.f. Habitat, Environment; *compare with* Stratum, social, human.

Stratum, social, human The human individual's social Stratum is originally determined by his parents and other relatives. His potential Stratum is mainly determined by his ontogenetic ceiling (q.v.). Hence he may move socially up or down during his lifetime. Thomas Jefferson distinguished the highest social Stratum as "artificial", the highest potential as "natural" aristocracy. These terms, however, are noisy. C.f. Noise, ontogeny, ceiling of; Stratum, potential.

Sub-strata (plural of *Stratum*, written with a capital S) The ontogenetic layers or levels in a system whose ceiling number (q.v.) determines the number of the Stratum in question. (*See* Strata). E.g., in plants, animals or persons the ontogenetic stage at the time in question. The number of its ontogenetic stages at maturity is its Stratum number (q.v.). See Number, characteristic.

Sub-stratum (n.) (singular of *Strata*) Ontogenetic stage (q.v.) of the individual in question, at the time in question: human, animal, plant, geoid, molecule, or atom. The individual's highest ontogenetic stage is his ontogenetic ceiling (q.v.). C.f. Ontogeny, stage of; *compare with* Ontogeny, ceiling of.

Supra-organism (term coined by Alfred E. Emerson ca. 1941) A pan-telic system composed of mono or di-telic individuals. E.g. a termite colony composed of so-called castes: worker, soldier and reproducers. C.f. Tel.

Symbiosis (n.) (Greek for living together) The coaction (q.v.) in which the system's work component and controller (q.v.) intensify each other's activity or goal achievement. Symbiosis, written $(+, +)$, is the characteristic coaction of Group IV in all Periodic Tables. Also called *co-operation, mutual help, give and take*.

Synnecrosis (n.) (Greek, dying together; coined term) The coaction (q.v.) in which the system's work component and controller (q.v.) decrease each other's activity. Synnecrosis, written $(-, -)$, is the characteristic coaction of Group VIII of all Periodic Tables. Synonym: *mutual harm, mutual ruin*, as of contending classes.

System, closed A non-existent theoretical system which receives no inputs (q.v.) and emits no outputs (q.v.). C.f. System, open.

System, Empirical Any concrete instance of the general system (q.v.), especially in which controller and work component (q.v.) are cybernetically related (q.v.); in which space-time boundaries are discernable, into which more inputs, and out of which outputs of energy or information (q.v.).

System, general (Defined by Haskell and Cassidy ca. 1950) A space-time region bounded by clear but not complete breaks of inter-dependence; the non-breaks consisting of inputs and outputs, and the inter-dependence consisting of causal and retroactive sequences between the system's basic components: work component and controller (q.v.).

System-hierarchy (coined term) A hierarchy of systems such that each member of the hierarchy after the first is composed of most or all preceding members plus one or more additional (emerged) structures and processes, mutually modified. E.g. the hierarchy of atoms from hydrogen to einsteinium. (W. V. Quine, H. G. Cassidy, E. Haskell, 1964).

System, open A system (q.v.) which receives inputs, emits outputs, or both. C.f. System, closed.

Systems Theory, General A body of systematical theoretical constructs which discuss the general relationships of the empirical world. (Kenneth Boulding).

Taxonomic System (of Unified Science) A classification of ecosystems (q.v.)—abiotic, biotic, and cultural—described in terms of their Characteristic numbers (q.v.).

Taxonomic System (of biology) A system of classification of plants and animals based upon the genetic descent of organisms, but without regard to their habitats (q.v.). This classification is a precondition for and is incorporated in the taxonomic system of Unified Science (q.v.). C.f. Characteristic numbers.

Technogeny (coined term) Forms of control (q.v.) such as manufacture, transport, agriculture, communication, simulation, etc. the capacity to do which is transmitted, not mentally, but genetically. E.g. nest construction by termites. C.f. Technology.

Technology The application of thought, primarily of scientific thought, to the solution of practical problems. Technology acts as a major empirical verifyer of scientific theory. (Figure V-1). C.f. Technology.

Tel (n.) (coined term) A system's goal or norm. (From the Greek *telos*, the end or goal; as in teleology). In ecology, the objectives of an organ, organism, or society. E.g. nutrition, reproduction, protection. A system may be monotelic, di-, tri- etc., to pantelic. E.g. a termite colony is pantelic, being divided into castes which are mono-telic, yet which fulfil all of the supra-organism's tels (q.v.).

Teleology (n.) Doctrine or view that developments are due to the purpose or design that is served by them. Hence implying a Designer. C.f. Teleonomy, Teleomorphy.

Telemorphy (n.) Displaying a form such as to contribute to or result in one or more systems higher than itself in the System-hierarchy (q.v.). C.f. Teleonomy, Teleology.

Teleonomy (n.) Displaying laws of behavior such as to contribute to or result in one or more systems higher than itself in the System-hierarchy (q.v.). C.f. Teleomorphy, Teleology.

Type specimen A paradigmatic case of a class or category of entities, such as a species of plant or animal; or of an ecosystem. A Periodic Table (q.v.) is just a theoretical framework to the extent to which its categories lack corresponding type specimens. (It is just as empirical system to the extent that its type specimens lack coherent theoretical explication.)

Value bias The general direction of a system's predominant coaction or moral value (q.v.), positive, neutral or negative. The Great Religions and Unified Science share positive value bias. Marxism and fascism share negative value bias. (E.g. "All of recorded history", says the *Communist Manifesto*, "is the history of class conflict"; conversely, "Love one-another," *New Testament*.)

Value, moral In Unified Science, the relation between any system's work component and controller (q.v.). This relation, also called *coaction* (q.v.), determines the system's properties, and is the criterion of its Group (q.v.) in the Periodic Table (q.v.). Geometrically, moral value is synonymous wit the radius vector's direction (θ). Coaction determines (and is determined by) the properties of the system, as stated by the General Periodic Law or Moral Law (q.v.).

Vector (n.) A symbol in geometry, esp. Periodic geometry (q.v.) in the form of an arrow whose length represents the quantity (q.v.) of the system in question and whose direction represents its quality (q.v.). C.f. Periodic Law, General.

Web of life The relationships, especially coactions (q.v.), among the members and components of an eco-system (q.v.). (See Figure II-14b). C.f. Web of mind.

Web of mind (coined term) The mental relationships, especially coactions (q.v.), among members of an institution or culture. (See Figures IV-5, IV-6.). C.f. Web of life.

Work Component The large and powerful component of a cybernetic system whose operation is influenced by the governor or controller. The less strategic entity in an ordinated relationship (q.v.).

Work Component The large and powerful component of a cybernetic system whose operation is influenced by the governor or controller. The less strategic entity in an ordinated relationship (q.v.). In the Periodic coordinate system (q.v.), the work component is mapped on the X axis. C.f. Controller, ordination, strategicity.

Zero, scalar (coined term) Synonym of *reference zero*, q.v.

Zero-zero circle (*See* Circle of atropy, Scalar zero circle, Reference zero; Atropy, circle of.

Zygote (n.) Fertilized ovum, the first ontogenetic stage or Substratum (q.v.) of higher organisms—plant, animal, or human.

The Authors

CASSIDY, Harold Gomes, born in Havana, Cuba, October 17, 1906; A.B. Oberlin College, 1930, A.M. 1932; Ph.D. (Chemistry) Yale, 1939. Member of Yale faculty, 1938–72, professor of chemistry, 1958–72. National Sigma Xi lecturer, 1960, 1965; Ayd lecturer, 1962; Korzybski Memorial lecturer, 1962; national lecturer, Scientific Research Society of America, 1965; senior fellow in science, Center for Advanced Studies, Wesleyan University, 1965–66; Danforth visiting lecturer, Association of American Colleges Arts Program, 1968, 1971; Sigma Xi centennial lecturer, Ohio State University, 1970. Recipient of third John Prymak service award, Connecticut Science Teachers Association, 1968; national award for excellence in chemistry teaching, Manufacturing Chemists Association, 1972. Fellow, American Association for the Advancement of Science, New York Academy of Sciences; secretary, Council for Unified Research and Education; member of New University Council. Author: (with J. English) *Principles of Organic Chemistry*, 1949; *Adsorption and Chromatography*, 1951; (with J. English) *Laboratory Book*, 1951; *Fundamentals of Chromatography*, 1957; *The Sciences and the Arts*, 1962; (with K. A. Kun) *Oxidation-Reduction Polymers*, 1965; *Knowledge, Experience and Action*, 1969; *Science Restated—Physics and Chemistry for the Non-Scientist*, 1970; numerous articles. Associate editor, American Journal of Science, 1948–67. Address: Hanover College, Hanover, Indiana.

CLARK, Jere Walton, born in Rex, Georgia, January 31, 1922; B.B.A. University of Georgia, 1947, M.A. 1949; Du Pont fellow, University of Virginia, 1949–51, Ph.D. (economics) 1953. Assistant professor, West Virginia University, 1952–55; associate professor, University of Chattanooga, 1955–62; professor of economics, Southern Connecticut State College, 1962– , chairman of Department of economics, 1966–70, director of Center for Interdisciplinary Creativity, 1967– . Recipient of award for best college course in economics, Calvin K. Kazanjian Economics Foundation, 1963. Chairman, task force on general systems education, Society for General Systems Research; executive director, Consortium on

Systems Education, New Haven. Address: Center for Interdisciplinary Creativity, Southern Connecticut State College, New Haven, Connecticut 06515.

HASKELL, Edward Fröhlich, born in Plovdiv, Bulgaria, August 24, 1906. A.B. Oberlin College, 1929; Columbia, 1929–30; Harvard, 1935–37; Chicago, 1937–40, fellow, 1940–43. Instructor in sociology (human, animal, plant) and anthropology, University of Denver, 1944, Brooklyn College, 1946–48. Chairman, Council for Unified Research and Education, 1948– . Consultant, West Virginia University, Drew University New School for Social Research, 1968; special lecturer, Southern Connecticut State College, Center for Interdisciplinary Creativity, 1969– . Fellow, American Association for the Advancement of Science, Society for Applied Anthropology; convening secretary, New University Council. Author: *Lance—A Novel About Multicultural Men*, 1941; (with Harold Cassidy) *Plain Truth and Redirection of the Cold War*, offset printed, 1961; (with a chapter by Harold Cassidy) *Unified Science*, Volume I, offset printed by National Institute of Health and xeroxed by IBM Systems Research Institute, 1969; various articles. Address: 617 West 113th Street, New York, N.Y. 10025.

JENSEN, Arthur Robert, born in San Diego, California, August 24, 1923. B.A. University of California at Berkeley, 1945; M.A. San Diego State College, 1952; Ph.D. (psychology) Columbia, 1956. Research fellow, Institute of Psychiatry, University of London, 1956—58; assistant professor of educational psychology, University of California at Berkeley, 1958–61, associate professor, 1962–66, professor, 1966– , associate resident psychologist, Center for Human Learning, 1961–66, resident psychologist, Institute for Human Learning, 1966– . Guggenheim fellow, Institute of Psychiatry, University of London, 1964–65; fellow, Center for Advanced Study in Behavioral Science, Stanford University, 1966–67. Author, *Genetics and Education*, 1972; numerous articles. Address: Department of Education, University of California, Berkeley, California 94720.

Editor's Acknowledgements

I have been helped with this book by far more of my colleagues, students and their friends, and in far more ways that I can list here; but not in more than I can acknowledge and thank you for.

Gil Roschini, Judy Culbertson and Cathy Bruno worked for several evenings on the alphabetical index. Judy repeatedly, after a hard day's work, typed parts of this book all night. Glenn Strait spent many days in New York editing, debating, and correcting technical details. In Paris, Claude typed and retyped the glossary for days. And Neil Winterbottom, abandoning his private pursuits for three months, made hundreds of astute corrections and improvements in the text, compiled the table of contents, the author's biographies, and assembled the alphabetical index. Then he flew to London and helped correct the proofs. During the weeks of this work his family entertained me most kindly and graciously.

Substantial grants were generously donated to expedite publication, so that it could precede the 1972 International Conference on Unified Science: by William L. Wallace, Senior Vice-President of the Olin Corporation, and by Farley Jones, President of the Unification Church, U.S.A. We, the participants from three continents, thank you for this.

And finally we acknowledge our debt to the authors whose works we have cited, quoted, or reproduced, and to their publishers listed in index and references—the intellectual, spiritual, moral, and technical giants who have put their shoulders to this work today and for past decades and centuries—our parents in the hierarchy of intellect.

Appendix I

TYPE SPECIMENS OR PARADIGM CASES

THE CONCEPT of type specimens or paradigm cases of theoretical constructs is commonplace in traditional astronomy, geology, biology and chemistry. In chemistry, pure samples of a chemical element listed in the Periodic Table of chemical elements, samples say of hydrogen or uranium, are type specimens or paradigm cases of the theoretical classes in question. In traditional biology, each species and sub-species of plants or animals listed in the taxonomic classification is keyed to one or more type specimens carefully preserved in cases, bottles, presses, and so forth, in various museums and other institutions. In astrophysics and geology, type specimens (paradigm cases of theoretical classes) are keyed to specific geoid bodies or geophysical entities.

There is, however, a strategic difference between the chemical classification and most others: The chemical classification includes both quantitative and qualitative, both Stratum-Period and Group characteristics, while the traditional biological and geoid classifications omit the latter. This happened because the atomic Groups are classes of coaction potentialities between the atom in question and its chemical habitat (classes of so-called chemical properties), whereas traditional taxonomic classifications are confined to entities and omit their habitats. They stop short of the next necessary step, classifying the coactions between them. Since coactions are traditionally called *moral* relations, it is their omission that deprives these traditional disciplines of moral force; and, of course, it is their inclusion which confers upon Unified Science and its organised disciplines their moral force (see Glossary). This will become very clear when we discuss human cultures, where moral relations have long been recognised.

The decisive advance displayed by ecology lies in its empirical study of, and emphasis upon, the coactions between the traditionally classified entities and their habitats. Unified Science contributes theoretical definition and classification of these coactions; their mapping into the framework called *Periodic Table* and the *Periodic Coordinate System*.

The first step, then, in defining type specimens in this domain, is to select ecosystems with clearly discernible controllers such as, say, beaver valleys, and to compile the most reliable written, pictorial, and film records of them in existence. The next step is to draw up the ecosystem's web-of-life, on the general pattern displayed in Figure II-14b. This involves clear assessment of each participant's Characteristic Number, relating each organismic and abiotic factor to the Periodic Table in each major respect. This will disclose role-duplications, triplications, and higher multiplications; reveal many unsuspected forms of indirect coaction; display gaps in our image of the web and lead us to fill them in. This procedure will yield us far more complete and detailed understanding of each participant's own ecosystem—its own organism habitat system—than we now have, and prepare the third step. This is to assess the quantities of the processes comprising the strategic organismic or abiotic factors' webs. This will, in turn, permit us to simulate strategic webs with a computer, modifying, introducing, or eliminating factors at will and predicting the consequences. This will permit empirical verification, with subsequent changes and improvements of our image (theory) until its accuracy is as great as necessary for feasible and effective control.

This project should coordinate large sectors of biological data and research, as the verification and completion of the Periodic Table of the Chemical Elements did for chemical data after 1869. This is a pre-condition for technology assessment.

Appendix II

CLASSIFICATIONS: QUANTITATIVE AND QUALITATIVE

THIS FRAMEWORK of the Periodic Table of Human Cultures can now be used to assemble some of the accumulated masses of anthropological data, much as the Periodic Table of Chemical Elements was used in 1869 for masses of data amassed by chemists. Such assembly begins most easily with the two aspects of natural classification: quantitative (R) on the table's left-hand side, and qualitative (θ) on its right-hand side.

These two aspects are displayed by the two traditional kinds of anthropological classifications. Large masses of anthropological data have been prepared for what J. S. Mill called *natural classification* by these preliminary classifications' famous authors.

Periodicity and Stratification (R) are displayed, for instance, by "The Material Cultures and Social Institutions of the Simpler Peoples" of Hobhouse, Wheeler and Ginsberg whose nomenclature has been incorporated into the Periodic Table. This huge classification is developed much further in Murdock's "Ethnographic Atlas". And these two classifications' most famous precursor was Lewis Morgan's classification of human cultures in "Ancient Society", 1877. While less detailed and precise, it went beyond them by including Period 5 (Literates) which Morgan called *Civilization*. The Periodic Table accommodates this Period, as also Periods 6 (Lower Industrialises), 7 (Higher Industrialists), and as many others as may become necessary.

Grouping (θ) is displayed in small, incipient classifications such as the comparative studies of values by Clyde Kluckhohn and others; and Ruth Benedict in *Patterns of Culture* among many others. These sets of abstract data were made operationally constant, and thus classifiable, by Ethel Albert's "Theory Construction for the Comparative Study of Values in Five Cultures". At any given time, she pointed out, every culture is what chemists call a *mixture*: its members display most or all of the theoretically possible values in various proportions. Nonetheless, every culture displays what Albert calls a *dominant* value premise by which all others, which she calls *deviant*, tend to be suppressed. Deviant values thus tend to be what physicists call *potential*. Under certain conditions, however, one or more of these deviant values may become acteal or *kinetic*, and tend to split the society; to generate conflicts within it. If it displaces the previously dominant value-premise in the society's controller, part or all of the culture is transmuted into another Group. By pointing out the role of the dominant values, Albert's theory thus permits qualitative classification of cultures and of their qualitative (Group) transmutations, past and future.

The Periodic Table of Human Cultures permits social scientists to relate these two kinds of classification to each other. It thus permits us to formulate and test the Moral Law, whose scientific version reads as follows: *The properties of human cultures are functions of their moral relations*. Stated in geometric terms this is, of course, the human case of the General Periodic Law,

$$R = f(\theta).$$

How can this law be tested in the psycho-social domain? In the same way as it has been tested in the physical domain. There, mixtures were carefully separated into their constituent elements. Mendeleev mapped these elements into the loci in the incipient Periodic table where they seemed to fit best in regard both to atomic weight and properties. In some cases, however, discrepancies appeared. So Mendeleev tested the accepted atomic weights (the analogue of coaction or dominant value bias as shown in the first two chapters) or the accepted properties; and, in a number of cases, he found one or both to be incorrect. *These corrections almost always confirmed the Periodic Law, thereby invalidating traditional misconceptions.*— Unified Science predicts that similar fitting and verifying of accepted properties and value-biases in the biological and psycho-social sciences will similarly confirm the Moral Law, and advance these sciences with similarly growing speed and power.

CULTURAL TYPE SPECIMENS

Let us illustrate this research program by means of two or three cultural type specimens or paradigm cases.

The Pueblo Indians' cultures—those of the Zuni, Hopi, Acoma—belonged clearly, at the time they were described by Ruth Benedict, to Period 4: Higher Agriculturalists. And their dominant value-premise was clearly cooperation; that is to say, Group IV (+, +). So this culture is to be mapped into the space provided for that particular kind of system.

Full Circle   247

At the opposite side of the Periodic coordinate system (though not of the Periodic Table) lies Group VIII $(-, -)$. (We cannot here consider specimens of its three Sub-Groups.) A type specimen of this Group was the culture called Dobu at the time in question. It belonged fairly clearly to Period 4, Higher Agriculturalists, and is also to be mapped into the space provided. Another type specimen of this Group appears to be the Kwakiutl culture of the Northwest coast of America at the time in question. You, the reader, can verify these statements by reading Benedict's description of the Kwakiutl and those of other antropologists. When last I inquired about the Kwakiutl, the Canadian official in charge wrote that the tribe had much less than a hundred members left. That sort of thing is very much what the $(-, -)$ dominant value bias would lead one to expect.

The next step would be to map a few score type specimens into the Periodic Table. But in doing so, it is important to take into account the so-called sighting errors to which one is most likely to be prone.

Anthropological Sighting

Classification of the many hundreds of human cultures, past and present, into these few Periodically repeated Groups, and prediction of their transmutations, can occur usefully to the extent that the principal variables peculiar to psycho-social sighting are recognized and compensated for. This may be called *anthropological sighting* on the analogy of tank, anti-aircraft or bomber sighting, where the velocities of the sighter of the target, and of wind must be ascertained, correlated, and compensated for.

One culture variable, already mentioned, is the degree of value mixture and of value dominance in each culture sample at a given time. The greater the dominance of one value-premise, and the others' corresponding absence or suppression at the time in question, the more clearly and conclusively the culture can be classified. The speed and direction (jointly called *velocity*) of a deviant value's emergence and challenge, determines the likelihood of the culture's transmutation: of its breakdown to a lower Period, its build-up to a higher Period, or its shift to a different Group in the same Period. Toynbee has expounded important laws of transmutation in Literate cultures (Period 5).—This variable is analogous to target velocity.

The second sighting variable, which is analogous to certain defects of the sighting mechanism, is the classifier's own mental-moral organization: his own grasp of the Periodic Law of Human Cultures. (In the history of chemistry, this mental variable produced the decisive difference between Newland's failure with his Octaves and Mendeleev's success with his Periodic Table, as described by Posin.—A famous example of this variation in anthropology is Ruth Benedict's own change of moral insight from 1934, when she copyrighted *Patterns of Culture*, to 1941 when she lectured on the same subject at Bryn Mawr College. In 1934 her sighting apparatus was cultural egalitarianism, which she and many others incorrectly labeled *cultural relativism*. It consists in flat contradiction of the Periodic Law. She stated this unequivocally in *Patterns of Culture's* concluding sentence, thus: "We shall arrive then at a more realistic social faith, accepting as grounds of hope and as a new basis for tolerance the coexisting and *equally valid* patterns of life which mankind has created for itself from the raw materials of existence" (italics added). This flat denial of the Moral Law spread widely, and has intensified the breakdown symptoms in our culture which Toynbee called *promiscuity*, *truancy*, and *drift*. This is the Existentialists' disastrous way of dealing with the multiplicity of sighting mechanisms which constitutes each individual culture's and sub-culture's compass; namely, the false assertion that they are *"equally valid"*; the abdication of responsibility of judging and assessing them relative to what Heisenberg calls *a central order*. (This subject is dealt with extensively in Chapter V.)

Frank Goble points out in *The Third Force* that by 1941, in her Bryn Mawr lectures, "Ruth Benedict was highly dissatisfied with the concept of cultural relativity, which was popular among anthropologists of her day, and with which her name has been closely associated. She struggled to develop a way of comparing various societies as unitary wholes or, in 20th-century terms, as "systems". She tried to correct her sighting error. Unfortunately, Benedict died in 1948. A part of her objective, however, was achieved by Ethel Albert a few years later, as shown above.

Benedict was nonetheless well on the way. Goble points out that "The terms that Ruth Benedict chose to describe the two types of society [which she had recognized as *least* equally valid] were 'high synergy' and low 'synergy.' The high synergy societies were those where people cooperated together for mutual advantage [Group IV societies, $(+, +)$]." Her type specimens were the Zuni (Pueblo Indians), Arapesh, Dakota, and Eskimo.

"Ruth Benedict described the bad societies as 'Societies with low social synergy where the advantage of one individual becomes the victory over another, and the majority who are not victorious must shift as they can' ". Her type specimens were Chuckchee, Ojibwa, Dobu, and Kwakiutl. Her sighting apparatus had clearly been rebuilt in conformity to the Moral Law, which is a long step in the directiion of the Periodic Table.

People, it seems, are endowed with moral ability in varying degrees, just as they are endowed with mathematical or linguistic ability. All the Great Religions, and above all the Christian, clearly display approaches to the Periodic Law. They are expressions of this moral sense in the terms of pre-Literate and Literate peoples, Period 5. With the emergence of Unified Science, the continuity of this moral ability's development in terms of Lower and High Industrial cultures, Periods 6 and 7, becomes clearly visible. The empirical sciences' three-century-long structural amorality, the detour which Arthur Koestler called "The Parting of the Ways", comes to an end as science comes *Full Circle*, merging C. P. Snow's *Two Cultures* and producing Walter Lippmann's long urged and hoped for *Public Philosophy* of Industrial civilization.

The third anthropological sighting variable is analogous to the sighter's or aimer's own velocity. This variable has been described in Chapter II, Section 7, and related to Einstein's sighting technique in physics by way of his free-falling elevator and rotating room analogies. There it was pointed out that highly autocratic or predatory cultures on one hand, strongly symbiotic cultures on the other, give rise in their inhabitants to strongly biased images of the world: People raised in the first tend to misinterpret cooperators as predators; people raised in the second tend to misinterpret predators as harmless cooperators.

To these and other ethno-centrisms should be added (or subtracted) the sighter's (ego-centric) temperament, his inborn tendency to distort his images of others in the direction of himself. Geometric classification of animal and human temperaments, and thus of this sighting variable, is presented in considerable detail in *Unified Science* Assembly of the Sciences Into a Single Discipline. This is done for human societies in our Chapter II: *The Coordinate System of Political Science*.

The other sciences differ from the psycho-social in the kinds of sighting errors they have to take into account. The relativity and indeterminacy principles in physics, for instance, belong to its sighting techniques. (Weighing and other measuring instruments supplement or correct sightig defects one level lower than the variables discussed here.) In all the sciences, the reduction of sighting defects and errors increases agreement among people in classification and communication.

Our Periodic Tables and the sighting techniques by which we decrease our sighting and classifying errors together constitute our model of Leibniz's Universal Characteristic.

Index

Ability: learning   157–58
　mental   159–62
　personality correlates of   162–63
Abstraction: ceilings   116, 117, 121, 129, 133, 138, 143
　levels of   115, 116, 119, 121, 125
Ackoff, Russell L.   92
Adams, John   132
Agriculturalists: lower, middle and higher   28
Allardyce, Gilbert   73, 76
Alpha   11, 23, 29, **34**, 36, 41, 44, 172, 173, 175, 176, 178, 180, 181, 189
American Association for the Advancement of Science   144, 200
Analysis: cybernetic   4, 5, 8
　moral   176
Anderson, O. Roger   27
Anomaly   67, 73, 76, 77
Anti-matter   37–39, 40, 41
Anti-Omega   37, 44
Aristocracy   126, 127
　artificial   66
　natural   74, 121, 132
Arthron   1
Atropy   44, 45, 51, 77
　axis of   7
　circle of   45, 48, 50

Bacon, Sir Francis   136
Baltzell, E. Digby   64, 66, 74, 124, 126, 127
Barbarians   126
Bateson, Gregory   63, 64
Becoming   176, 183
　Sovereign   169, 180
Beer, Stafford   148, 149, 154
Bertallanffy, Ludwig von   47
Biology   52, 53
Biopoesis   26, 50
Bitterman, M. E.   107, 157

Black holes   37
Bohr, Niels   52, 54
Bolsheviks   56
Bonhoeffer, Dietrich   187
Bonaparte, Napoleon   66
Bourgeoisie   65–66

Campbell, Ernest   209–10
Campbell, Joseph   203
Capitalism   152, 192–93
Cassidy, Frederick G.   1
Cassidy, Harold G.   1–2, 37, 49, 50, 54, 55, 73, 180, 196, 200
Caste   126, 127
Cattell, R. B.   163
Causation   32
Celsius   39
Center, political: Creative   75, 135
　in Etats-Généraux   65
Central Order   43, 44, 118, 137, 172, 189
Centrism: logo-   72, 74, 77, 90
　patho-,   72, 74, 77
Centrists   77
Chance: evolution by   22
Chemical Society   56
Christianity   204–5
Circle: of Atropy   45, 48, 50
　of Perfectibility   137, 144
　of Perfection   135–37
　Reference Zero   6–7, 14, 16
　Relative Zero   39, 44
　Scalar Zero   35, 36, 41
　Zero-Zero   6–7, 35
City College, New York   133, 141
Civilization: breakdown of   63
　disintegration of   63, 75
　genesis of   55, 63, 64, 129
　Graeco-Roman   34
　higher industrial   126, 149, 188
　literate   63, 178, 198
　lower industrial   68, 118, 128, 198

Clark, Jere W. 89–90
Clarke, Gerald 203
Class: conflict 67, 69, 71
  cooperation 66, 69, 71, 77, 123, 191, 192
  new 66, 67, 132
  social 116
Classification: cybernetic 8–10
  ecological 54, 62
  natural 199
  taxonomic 8, 62
Cline: eco- 119
  geo- 120
  lingua- 119–20
Club of Rome 186, 200
Coaction cardioid 48
Coactions 5–7, 32, 198
  abiotic 45–46, 49–52
  animal 46
  atomic 46, 49–52
  biotic 45–46, 52–62
  cultural 45–46, 62–69
  negativization of 126
  objective 80, 143
  positive 75, 189
  subjective 80
Coaction vector 14
Communication 4, 12
  fouling 65
  medium of 99–100
Communism 68, 76, 195
Compass 124, 172, 173
  coaction 43, 65, 78, 211
  moral 43
Complexity: organized 11, 24
Computer simulation 55, 59, 117, 124, 175
Conant, James B. 65, 66, 144
Conflict: class, national and race 71
Conflictor's deficit 7
Consciousness 43
Conversation 76
Control 4–5, 12
  decline of 126
Controller (of system) 4–5, 32, 46, 59, 67
Cooperator's surplus 7
Coordinate system: Alpha 38–39, 40, 41
  Cartesian 15, 32–34, 36, 40–41, 48
  hyper-spatial 40, 41
  Inverted Periodic 35, 37–38, 40
  of political science 69–78, 107
  Periodic 6–7, 13, 32–37, 39, 40, 43, 47–48, 74, 76, 152, 153, 171–73, 175
Copernicus, Nicholas 108, 186–87
Correspondence 135
  principle of 52–53
Council for Unified Research and Education 199–200
Crisis prevention 93
Cromwell, Oliver 64
Cultures: autocratic 72
  democratic 72
  agricultural 28
  higher industrial 173, 177, 194
  hunting and gathering 28
  literate 28, 63
  lower industrial 28, 177
  pastoral 28
Cup of Life 28, 35, 36, 43
Cybernetics 4–5, 100, 124
  eco- 98–99

De Latil, Pierre 47
Democracy: breakdown of 133
  egalitarian 133
  and derangement of the governing power 133–35
  totalitarian 66
Descartes, René 34, 36, 40
Devolution 29, 36
Disintegration: rhythm of 63, 127
Djilas, Milovan 66, 67
Dropping out 139
Dutt, R. Palme 75
Duttweiler, Gottlieb 64, 192, 194

Eblen, William 45, 97
Ecology 144–45, 146
  general 98
  ecological roles 58
Ecosystem 54, 61
  animal 58
  human 124
  plant 57
Ectropy 28, 44, 48, 174, 177
  law of 52
Education: American 129–30
  Unified Science and 12–13
  Soviet 130–33
Egalitarianism, anthropological 74
Einstein, Albert 28, 50, 72
Electron: energy-level structure 7–8, 24
  shells 49–52

Index

Emergence 21, 29
Endogamy 118–19
Ensombrement 136, 137
Entropy 11, 28, 44, 48, 51, 174
Environmentalism 74
Equalization of opportunity 138–44
Etats-Généraux 65
Ethics 176
Evil 44, 50, 76, 77, 173, 174
Evolution 9, 11, 29
  human 115
  structure of 136
  theory of 62
Existentialism 180
Extremism 76, 77, 107
Eysenck, H. J. 162

Fascism 68, 73, 76
Fiellin, Alan 141, 143
Fosdick, Harry Emerson 204, 205
Fouling: signal and communication 53, 65, 76
Franklin, Benjamin 76
Freedom 137
Front: horizontal 64, 67, 195
  vertical 64, 65, 66, 67, 192, 193, 194
Gause, G. F. 13, 14, 36
Generalist 2, 104, 146, 149, 154–56
Genetics 62
  and mental development 159–62
Genotype: human 116, 117, 125
Geometry: Cartesian 32, 34, 36, 38, 41
  Periodic 6–7, 32–37
Gibbs, Willard 51
Gibbs' Law 51
God 28, 43, 50, 79, 177, 189
Good 43, 44, 50, 77, 173, 174
Government: derangements of 133–37
Groups: of chemical elements 31, 47, 49–52
  of general systems 32, 47, 51, 77

Habitat 24
  control by 9
  control of 9
  educational 138
  socio-cultural 116, 117, 121
Habit reversal 158
Harlow, H. F. and M. K. 158
Harr, K. G. 95
Haskell, Edward 1, 3, 4, 6, 8, 11, 12, 95, 96, 98, 101, 157, 158

Heisenberg, Werner 43, 44, 45, 50, 78, 172
Hierarchy: of foods 111, 114
  of social Strata 114, 121
  of social Sub-strata 111
  of psycho-genetic structures 114
  of tools 111, 114
  of vocabularies 111, 114
  Systems- 7–11, 21, 47, 64, 67, 69, 124, 136, 169, 176, 180, 181, 182–83
Hitler, Adolf 76
Hooke, Robert 199
House of Commons 65
Hunters: lower and higher 28
Hutchinson, G. Evelyn 36
Huxley, Sir Julian 92

Ideologies 64, 68, 72, 74, 140, 177
Ideologists 67, 68, 72, 136, 140
Idol of the Tribe 136
Ignorance explosion 149
Induction: problem of 182–83
Industrialists: higher 77, 123, 144, 177
  lower 28, 118, 125, 138, 140, 144, 198
Inequality: creative 75
Information 4, 5
Input 4
Inscrutability of reference 181
Intelligence: abstract 163
  animal 107
  human 107
  moral 163
  testing of 131, 132, 133, 139, 140
Invariance 3
Invariant relations 3, 11, 180, 187, 188
Iron curtain 72
Isomorphism 68

James, William 197
Jefferson, Thomas 74, 89, 132
Jensen, Arthur R. 107–9, 117
Jesus 204
Jonas, Hans 169, 171, 173–77
Judge, Anthony J. N. 93

Kant, Immanuel 69
Kelvin 39
Kendall, Henry W. 141
Kenyatta, Yomo 138
Koestler, Arthur 68, 197, 198
Kohlberg, L. 232
Kolers, Paul A. 102

Kristol, S.   152
Kuhn, Thomas   53, 67, 153, 186

Landring of the Independents   63, 194, 195
Langer, Suzanne   52
Langlois, T. H.   46, 47, 52, 62
Language: background   183, 184
  meta-   56, 59, 73, 92–93, 99–104, 149
Laplace   34
Law: moral   43–46, 48, 62, 68–69, 74, 77, 80, 174, 189
  natural   62, 69, 80, 136, 189
  Periodic   46, 48, 62, 64, 68–69, 71, 198
  universal   48
Leadership procedure sequence   200
Learning   157, 160
  1.-to-learn   158
Left (wing)   65, 73, 74
  far and extreme   68, 71, 74, 75, 76, 135
Leibniz, Gottfried Wilhelm von   41, 48, 55, 78–79, 80, 123, 190, 196
Lenin, V. I.   191
Life   11
  web of   55, 56, 57–59, 79
Lincoln, Abraham   128
Lippmann, Walter   126, 130, 131, 133–35, 136, 177, 178–79, 182, 188
Livesey, L. J.   80
Lodge, George C.   149
Lodge, George T.   175
Logic   182, 183
Loss of legality   75
Lorentz Transformation   3, 180
Love   75
LöwenheimöSkolem theorem   185
Lunt, Paul   127

Madden, Carl H.   152–53
Magna Carta   64
Majority   47
  external   126, 127
  internal   125, 126
  in Etats-Généraux   65–66
  number of social Strata in   112, 114, 125
  silent   67
Management   150, 151, 154
Mandate of Heaven   179
Marx, Karl   191, 201
Marxists   190, 191, 195
McElroy, William   27

McNeill, William   186
Mendeleev, D. I.   31, 32, 46, 47, 49, 54, 56, 58, 62, 69, 71, 102, 112, 198
Migros   63, 64, 192–95
Mill, John Stuart   138
Mind: human   43, 182, 183, 189
  web-of-   137, 141
Minority   47, 62, 67
  creative   3, 64, 66, 194
  dominant   663, 64, 66, 194
  number of social Strata in   114, 125
Moon, Sun Myung   187, 208
Moss, N. Henry   104
Multiversity   136, 14–49, 171
Mussolini, Benito   73, 76
Mutations: human   116, 121
  pleiotropic   115
Myths   202, 203

National Academy of Science   140
National Science Foundation   27
Natural empire   24, 115, 144
Natural kingdom   21, 41, 118
  of Animals   8, 27–28
  of Human Cultures   11, 23, 27, 28, 121
  of Plants   8–10
Natural kinds   41
Nature   176
Nerve structure   158
New ideology   155
Newton, Sir Isaac   40
New York Academy of Sciences   131
Nicholson, Marjorie Hope   79, 135–36
Nietzsche, Frederick   169
Noösphere   28
Northrop, F. S. C.   124
Numbers: characteristic   54–57, 58, 59, 61, 62, 115, 118, 124, 181, 182
  Natural   30, 34, 185
  negative   34, 36, 40, 41
  positive   34, 40

O'Connor, Johnson   115, 119
Omega   11, 24, 36, 44, 172, 173, 175–78, 189, 209
Ontogeny: of atom population   52
  of human mental abilities   158–59
Open admission   133, 141, 143, 144
Organization: levels of   111
  mental   44
Ostension, problem of   181–82
Output   4

Index

Palaeontology 144, 146
Palingenesia 178
Panofsky, Wolfgang K. H. 141
Paradigm 68, 73
  of cultural relativism 69
  of existentialism 69
  of political theory 66, 67
  of Unified Science 68, 69, 109, 154, 180, 188–89
  system of 154
Parasitism 62
Pasteur, Louis 108
Pastoralists: lower and higher 28
Pauli, Wolfgang 78
Pauling, Linus 50, 51
Peccei, Aurelio 186, 200
Pennsylvania 76
Periodic table 1, 3, 55
  of animal ecosystems 59–62
  of chemical elements 4, 8, 30–32, 49–52, 56, 111, 114
  of human cultures 111–16, 121–23, 124
  of plant ecosystems 62
  of stable particles 38
Periodicity 4, 7, 52
Periods 4, 8, 41
  animal, human and plant 36
  atomic 8, 49–52
  major 24
Petrie, M. Ann 141, 142, 143
Phenotype: human 116, 121
Philosophy: public 135, 137, 149, 152, 154, 177, 178, 179
  solid 200
Phylogeny of adaptive behavior 157–58
Physics: Newtonian and Einsteinian 53
Piaget, J. 158
Pinkham, Rossalie 97
Plato 172
Pleiotropic: characters 161
  gene-complex 121
  mutations 115
Predation 62, 64, 127
Problem-solving 94–95
Process 4–5
  homeostatic 5
Proletariat 66, 67, 125
  vanguard of the 67
Proxy function 184
Psycho-genetic tree 114–15, 121, 123

Quasars 24–26, 51

Quine, Willard V. 41, 55, 61, 180–83, 185

Race: conflict 69, 71
  cooperation 77
Racism 140
Radius space vector 48, 59
Rate-change: geometrization of 48
Reference frame 11, 12, 71
  Newtonian 72–73
  political 71, 74, 75, 77
Relations: cultural 45, 46
  cybernetic 32
  invariant 3, 11, 180, 187, 188
  moral 29
  social and political 62
Relativism: cultural 69, 74
Relativity 3
  ontological 41, 180
  physical 41
  theory of 71
Religions 27, 66, 68, 69, 71, 177, 202–3
Revolutions: French and Russian 64
  scientific 53, 77, 152, 153, 187, 189
Right (wing) 65, 73, 74
  far and extreme 68, 71, 75, 76, 135
Robinson, John A. T. 71, 187
Royal Society 56, 197, 199

Scanlon Plan 7
Schizm: of the body politic 63, 178
  of the soul 63, 126, 127, 178, 186
Schizmogenesis 63, 64, 66
Schools for phisosopher-kings 131–32
Science: behavioral 156
  one-field 41, 44, 169, 170–71, 180
  political 64, 73, 76, 117
  psycho-social 72, 77, 140, 149
  traditional structure of 99
  unified 51, 73, 77, 95–98, 106, 124, 137, 150, 151, 156, 169, 170, 172–79, 183, 197
Scientific Monthly 43
Seaborg, Glenn T. 30, 51, 58
Second International Congress for the Philosophy of Science 72
Sense of similarity 182, 183
Snow, C. P. 151, 187–88
Social mobility 116–17, 118, 121, 161
Social Strata 116–21
Socialism 152, 153
Social capitalism 152, 153, 192, 194

Society: classless   123, 132
  industrial   63
  literate   63
  lower agricultural   63
  new   93
Society for General Systems Research   130
Sonnemann, Ulrich   144–48
Specialist   2, 146
  as a psychological problem   145–48
  illegitimate   146, 147
  legitimate   146
  organized   145, 149
System: breakdown and disintegration of   117
  closed   11
  controller of   4, 5, 32
  general   178
  meta-   149
  natural   32, 188
  open   11
  two-ideology   71, 72, 75, 76
  two-party   65, 70
  work component of   4, 5, 32, 114

Talmon, J. L.   66
Technogeny   28
Technology   28
Teilhard de Chardin, Pierre   11, 28, 187, 209
Teleology   21, 189
Teleomorphy   21, 189
Teleonomy   21, 46, 69, 189
Temperaments   194
Tests (of ability)   131, 132, 133, 139, 140, 144
Theoreticians   67, 107
Theory: background   2, 55, 73, 77, 150, 180
  coaction   141
  general systems   124, 178
  of evolution   199
  socio-political   52, 66–67, 77
Thermodynamics: Second Law of   11
Tillich, Paul   187
Tocqueville, Alex de   127
Toffler, Alvin   175
Toynbee, Arnold J.   28, 47, 55, 62–63, 68, 71, 124, 125, 126, 178, 198
Transfiguration (of society)   178

Transmutation of society   117, 121, 123, 177, 190, 192, 198, 204
Twain, Mark   89, 101
Two Cultures   40, 46, 151, 187–88
Type specimens   61, 204

Unification Church   206, 207
Universal Characteristic   41, 79, 80, 123, 153, 196
University   128–29, 149, 154
U.S. Atomic Energy Commission   30

Value bias: negative   71, 180
  positive   71, 74–75, 178, 189, 203
  zero   71, 180
Value premise: deviant   123
  dominant   66, 71, 123
  negative   68, 71
  positive   68, 178
  zero   68
Values   43, 45–46, 68
  objective   69, 71
  one-field science and   68, 71
  subjective   71
  unified science and   45–46, 68
Vector-direction   173
Vocabulary: ceiling   115
  coaction-   53
  levels of   116, 119, 121, 142, 143

Warner, W. Lloyd   118, 124, 126, 127
Web: of Life   55, 59, 79, 181
  of Mind   137, 141
Weaver, Warren   24, 44, 52
Wells, H. G.   127
White, Sheldon   158
Whitney, Eli   89
Wiener, Norbert   4, 47, 138
Williams, George L.   126
Witt, A. A.   36
Work component   4, 5, 32, 46, 59
World War III   68
Wu, C. S.   40

Zerequal axis   15
Zero   34
  absolute   39
  relative   39
Zygotes   62